DIANE ACKERMAN'S
A NATURAL HISTORY OF THE SENSES

"This is one of the best books of the year—by any measure you want to apply. It is interesting, informative, very well written. This book can be opened on any page and read with relish....thoroughly delightful...Don't miss it." —*St. Petersburg Times*

"This book is pure ecstasy. It is a treasure trove of information, diverse in space and time and culture but all related to the pleasures of sensory experience." —*Houston Chronicle*

"Ms. Ackerman is an athlete of the senses....To think our way back into feeling: this is [her] mission, and she's very persuasive. On every other page, there's a nice aperçu." —*The New York Times Book Review*

"[Ackerman's] fascinating book inspires an enthusiasm for the diversity of human experience and is a tribute to the amazing power of our senses. It's both a sensual feast and a celebration." —*Seattle Times*

"*A Natural History of the Senses* is as voluptuous a volume as its subject matter cries out for. The charm of Diane Ackerman's book is that it arouses awareness and appreciation of sensual life. In small, tasty morsels, it will delight you." —*Los Angeles Times Book Review*

"An intriguing, knowledgeable and compelling book on the science, mood, character and geography of the human senses. But...it is [Ackerman's] inquiry into the temper and disposition of the senses that endures and settles irresistibly just beneath the reader's skin. In exploring the extreme diversity of the human senses and their incredible variegation from culture to culture, Ms. Ackerman manages to reveal just how exceptional, rather than common, human senses are." —*Atlanta Journal and Constitution*

"Often funny, often poignant...The synthesis here—Ackerman's ability to help us see that the sum of our senses is greater than the individual parts, and to do so in language that often resembles a prose poem—is all the more impressive for her finesse in linking science with our loftier aspirations." —*San Francisco Chronicle*

ALSO BY DIANE ACKERMAN

The Moon by Whale Light

Jaguar of Sweet Laughter

Reverse Thunder

On Extended Wings

Lady Faustus

Twilight of the Tenderfoot

Wife of Light

The Planets: A Cosmic Pastoral

A NATURAL HISTORY OF THE SENSES

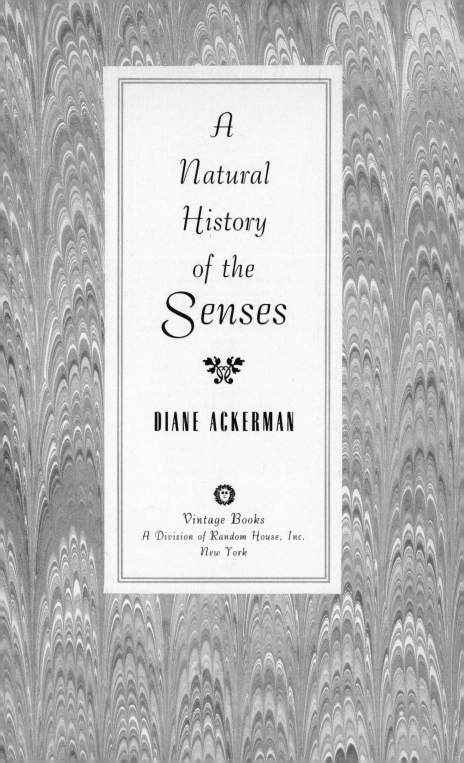

A
Natural
History
of the
Senses

DIANE ACKERMAN

Vintage Books
A Division of Random House, Inc.
New York

First Vintage Books Edition, September 1991

Portions of this work were originally published
as first-serial contributions to
Parade magazine. Portions of this work were originally
published in different form in
The New York Times Book Review
and *Condé Nast Traveler*.

Owing to limitations of space,
all acknowledgments for permission to reprint
previously published material
may be found on page vii.

Library of Congress Cataloging-in-Publication Data
Ackerman, Diane.
A natural history of the senses / Diane Ackerman. —1st Vintage
Books ed.
p.   cm.
Includes index.
ISBN 0-679-73566-6
1. Senses and sensation.   2. Manners and customs.   3. Human
behavior.   I. Title.
[BF233.A24   1991]
152. 1—dc20   91-50048
CIP

Book design by Debbie Glasserman

Manufactured in the United States of America
10 9 8 7

The initial mystery that attends any journey is: how did the traveller reach his starting point in the first place? How did I reach the window, the walls, the fireplace, the room itself; how do I happen to be beneath this ceiling and above this floor? Oh, that is a matter for conjecture, for argument pro and con, for research, supposition, dialectic! I can hardly remember how. Unlike Livingstone, on the verge of darkest Africa, I have no maps to hand, no globe of the terrestrial or the celestial spheres, no chart of mountains, lakes, no sextant, no artificial horizon. If ever I possessed a compass, it has long since disappeared. There must be, however, some reasonable explanation for my presence here. Some step started me toward this point, as opposed to all other points on the habitable globe. I must consider; I must discover it.

—Louise Bogan, *Journey Around My Room*

A mind that is stretched to a new idea never returns to its original dimension.

—Oliver Wendell Holmes

PERSONAL ACKNOWLEDGMENTS

Many friends and acquaintances have sent me useful books and articles, or shared reminiscences with me about the senses. I'm indebted especially to Walter Anderson, Ronald Buckalew, Whitney Chadwick, Ann Druyan, Tiffany Field, Marcia Fink, Geoff Haines-Stiles, Jeanne Mackin, Charles Mann, Peter Meese, the Monell Chemical Institute, Joseph Schall, Saul Schanberg, Dava Sobel, Sandy Steltz, and Merlin Tuttle. My special thanks to Dr. David Campbell and Dr. Roger Payne, who were generous enough to cast an eye over the manuscript, looking for infelicities.

Almost every week, a familiar buff-colored envelope would arrive from my editor, Sam Vaughan, whose leads, suggestions, and questions I grew to rely on, and whose friendship I've come to cherish.

Parade magazine first published four excerpts from "Touch," "Vision," and "Smell."

"Courting the Muse" appeared in *The New York Times Book Review*. Part of "Why Leaves Turn Color in the Fall" appeared in a different form in *Condé Nast Traveler*.

"How to Watch the Sky" was initially prepared for the National Geographic Society's book *The Curious Naturalist* and is reproduced here with my gratitude for their understanding.

CONTENTS

INTRODUCTION

IN EVERY SENSE XV

TASTE 125

HEARING 173

VISION 227

Introduction

IN EVERY SENSE

How sense-luscious the world is. In the summer, we can be decoyed out of bed by the sweet smell of the air soughing through our bedroom window. The sun playing across the tulle curtains gives them a moiré effect, and they seem to shudder with light. In the winter, someone might hear the dawn sound of a cardinal hurling itself against its reflection in a bedroom windowpane and, though asleep, she makes sense enough of that sound to understand what it is, shake her head in despair, get out of bed, go to her study, and draw the outline of an owl or some other predator on a piece of paper, then tape it up on the window before going to the kitchen and brewing a pot of fragrant, slightly acrid coffee.

We may neutralize one or more of our senses temporarily—by floating in body-temperature water, for instance—but that only heightens the others. There is no way in which to understand the world without first detecting it through the radar-net of our senses. We can extend our senses with the help of microscope, stethoscope, robot, satellite, hearing aid, eyeglasses, and such, but what is beyond our senses we cannot know. Our senses define the edge of consciousness, and because we are born explorers and questors after the unknown, we spend a lot of our lives pacing that windswept perimeter: We take drugs; we go to circuses; we tramp through jungles; we listen to loud music; we purchase exotic fragrances; we pay hugely for culinary novelties, and are even willing to risk our lives to sample a new taste. In Japan, chefs offer the flesh of the puffer fish, or *fugu*, which is highly poisonous unless prepared with exquisite care. The most distinguished chefs leave just enough of the poison in the flesh to make the diners' lips tingle, so that they know how close they are coming to their mortality. Sometimes, of course, a diner comes *too* close, and each year a certain number of *fugu*-lovers die in midmeal.

How we delight our senses varies greatly from culture to culture (Masai women, who use excrement as a hair dressing, would find American women's wishing to scent their breath with peppermint equally bizarre), yet the way in which we use those senses is exactly the same. What is most amazing is not how our senses span distance or cultures, but how they span time. Our senses connect us intimately to the past, connect us in ways that most of our cherished ideas never could. For example, when I read the poems of the ancient Roman poet Propertius, who wrote in great detail about the sexual response of his ladyfriend Hostia, with whom he liked to make love by the banks of the Arno, I'm amazed how little dalliance has changed since 20 B.C. Love hasn't changed much, either: Propertius pledges and yearns as lovers always have. More remarkable is that her body is exactly the same as the body of a woman living in St. Louis right now. Thousands of years haven't changed that. All her delicate and quaint little "places" are as attractive and responsive as a modern woman's. Hostia may have interpreted the sensations differently, but the information sent to her senses, and sent by them, was the same.

If we were to go to Africa, where the bones of the petite mother of us all, Lucy, lie, just where she fell millennia ago, and look out across the valley, we would recognize in the distance the same mountains she knew. Indeed, they may well have been the last thing Lucy saw before she died. Many features of her physical world have changed: The constellations have shifted position a little, the landscape and weather have changed some, but the outlines of that mountain still look much the same as when she stood there. She would have seen them as we do. Now leap for a moment to 1940 in Rio de Janeiro, to an elegant home owned by the Brazilian composer Heitor Villa-Lobos, whose music, both rigorous and lavish, begins with the tidy forms of European convention and then explodes into the hooting, panting, fidgeting, tinkling sounds of the Amazon rain forest. Villa-Lobos used to compose at the piano in his salon—he would open the windows onto the mountains surrounding Rio, choose a vista for the day, draw the outline of the mountains on his music paper, then use that drawing as his melodic line. Two million years lie between those two observers in Africa and Brazil—their eyes making sense of the outline of a mountain—and yet the process is identical.

The senses don't just *make sense* of life in bold or subtle acts of clarity, they tear reality apart into vibrant morsels and reassemble them into a meaningful pattern. They take contingency samples. They allow an instance to stand for a mob. They negotiate and settle for a reasonable version and make small, delicate transactions. Life showers over everything, radiant, gushing. The senses feed shards of information to the brain like microscopic pieces of a jigsaw puzzle. When enough "pieces" assemble, the brain says *Cow. I see a cow.* This may happen before the whole animal is visible; the sensory "drawing" of a cow may be an outline, or half an animal, or two eyes, ears, and a nose. In the flatlands of the Southwest, a speck develops a tiny line at the top. *Cowboy,* the brain says, a person who has turned his head, revealing the silhouette of a hat brim. Sometimes the information arrives second- or thirdhand. A roll of dust in the distance: a pickup truck at speed. *Reasoning* we call it, as if it were a mental spice.

A sailor stands on the deck of a ship, holding semaphore flags snug against his side. Suddenly he lifts them, swings both to the right in a take-it-away-man gesture, then turns, squats and sweeps the flags overhead. The sailor is a sense transmitter. Those who see and read him are the receptors. The flags are always the same, but how he moves them differs depending on the message, and his repertoire of gestures covers many contingencies. Change the image: A woman sits at a telegraph key and rattles Morse code along a wire. The dots and dashes are nerve impulses that can combine in elaborate ways to make their messages clear.

When we describe ourselves as "sentient" beings (from Latin *sentire,* "to feel," from Indo-European *sent-,* "to head for," "go"; hence to go mentally) we mean that we are conscious. The more literal and encompassing meaning is that we have sense perception.

"Are you out of your senses!" someone yells in angry disbelief. The image of someone sprung from her body, roaming the world as a detached yearning, seems impossible. Only ghosts are pictured as literally being out of their senses, and also angels. *Freed* from their senses is how we prefer to say it, if we mean something positive—the state of transcendental serenity found in an Asiatic religion, for example. It is both our panic and our privilege to be mortal and sense-full. We live on the leash of our senses. Although they enlarge us, they also limit and restrain us, but how beautifully. Love is a beautiful bondage, too.

We need to return to feeling the textures of life. Much of our experience in twentieth-century America is an effort to get away from those textures, to fade into a stark, simple, solemn, puritanical, all-business routine that doesn't have anything so unseemly as sensuous zest. One of the greatest sensuists* of all time—not Cleopatra, Marilyn Monroe, Proust, or any of the other obvious voluptuaries— was a handicapped woman with several senses gone. Blind, deaf, mute, Helen Keller's remaining senses were so finely attuned that when she put her hands on the radio to enjoy music, she could tell the difference between the cornets and the strings. She listened to colorful, down-home stories of life surging along the Mississippi from the lips of her friend Mark Twain. She wrote at length about the whelm of life's aromas, tastes, touches, feelings, which she explored with the voluptuousness of a courtesan. Despite her handicaps, she was more robustly alive than many people of her generation.

We like to think that we are finely evolved creatures, in suit-and-tie or pantyhose-and-chemise, who live many millennia and mental detours away from the cave, but that's not something our bodies are convinced of. We may have the luxury of being at the top of the food chain, but our adrenaline still rushes when we encounter real or imaginary predators. We even restage that primal fright by going to monster movies. We still stake out or mark our territories, though

---

*Someone who rejoices in sensory experience. A sensualist is someone concerned with gratifying his sexual appetites.

sometimes now it is with the sound of radios. We still jockey for position and power. We still create works of art to enhance our senses and add even more sensations to the brimming world, so that we can utterly luxuriate in the spectacles of life. We still ache fiercely with love, lust, loyalty, and passion. And we still perceive the world, in all its gushing beauty and terror, right on our pulses. There is no other way. To begin to understand the gorgeous fever that is consciousness, we must try to understand the senses—how they evolved, how they can be extended, what their limits are, to which ones we have attached taboos, and what they can teach us about the ravishing world we have the privilege to inhabit.

To understand, we have to "use our heads," meaning our minds. Most people think of the mind as being located in the head, but the latest findings in physiology suggest that *the mind* doesn't really dwell in the brain but travels the whole body on caravans of hormone and enzyme, busily making sense of the compound wonders we catalogue as touch, taste, smell, hearing, vision. What I wish to explore in this book is the origin and evolution of the senses, how they vary from culture to culture, their range and reputation, their folklore and science, the sensory idioms we use to speak of the world, and some special topics that I hope will exhilarate other sensuists as they do me, and cause less-extravagant minds at least to pause a moment and marvel. Inevitably, a book such as this becomes an act of celebration.

A NATURAL HISTORY OF THE SENSES

Smell

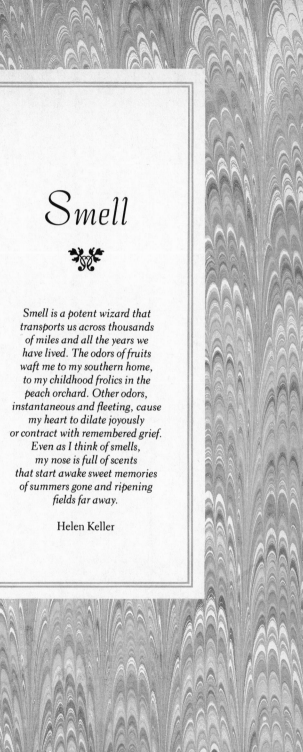

Smell is a potent wizard that
transports us across thousands
of miles and all the years we
have lived. The odors of fruits
waft me to my southern home,
to my childhood frolics in the
peach orchard. Other odors,
instantaneous and fleeting, cause
my heart to dilate joyously
or contract with remembered grief.
Even as I think of smells,
my nose is full of scents
that start awake sweet memories
of summers gone and ripening
fields far away.

Helen Keller

THE MUTE SENSE

Nothing is more memorable than a smell. One scent can be unexpected, momentary, and fleeting, yet conjure up a childhood summer beside a lake in the Poconos, when wild blueberry bushes teemed with succulent fruit and the opposite sex was as mysterious as space travel; another, hours of passion on a moonlit beach in Florida, while the night-blooming cereus drenched the air with thick curds of perfume and huge sphinx moths visited the cereus in a loud purr of wings; a third, a family dinner of pot roast, noodle pudding, and sweet potatoes, during a myrtle-mad August in a midwestern town, when both of one's parents were alive. Smells detonate softly in our memory like poignant land mines, hidden under the weedy mass of many years and experiences. Hit a tripwire of smell, and memories explode all at once. A complex vision leaps out of the undergrowth.

People of all cultures have always been obsessed with smell, sometimes applying perfumes in Niagaras of extravagance. The Silk Road opened up the Orient to the western world, but the scent road opened up the heart of Nature. Our early ancestors strolled among the fruits of the earth with noses vigilant and precise, following the seasons smell by smell, at home in their brimming larder. We can detect over ten thousand different odors, so many, in fact, that our memories would fail us if we tried to jot down everything they represent. In "The Hound of the Baskervilles," Sherlock Holmes identifies a woman by the smell of her notepaper, pointing out that "There are seventy-five perfumes, which it is very necessary that a criminal expert should be able to distinguish from each other." A low number, surely. After all, anyone "with a nose for" crime should be able to sniff out culprits from their tweed, India ink, talcum powder, Italian leather shoes, and countless other scented paraphernalia. Not to mention the odors, radiant and nameless, which we decipher without even knowing it. The brain is a good stagehand. It gets on with its work while we're busy acting out our scenes. Though most people will swear they couldn't possibly do such a thing, studies show that both children and adults, just by smelling, are able to determine whether a piece of clothing was worn by a male or a female.

Our sense of smell can be extraordinarily precise, yet it's almost impossible to describe how something smells to someone who hasn't smelled it. The smell of the glossy pages of a new book, for example, or the first solvent-damp sheets from a mimeograph machine, or a dead body, or the subtle differences in odors given off by flowers like bee balm, dogwood, or lilac. Smell is the mute sense, the one without words. Lacking a vocabulary, we are left tongue-tied, groping for words in a sea of inarticulate pleasure and exaltation. We see only when there is light enough, taste only when we put things into our mouths, touch only when we make contact with someone or something, hear only sounds that are loud enough. But we smell always and with every breath. Cover your eyes and you will stop seeing, cover your ears and you will stop hearing, but if you cover your nose and try to stop smelling, you will die. Etymologically speaking, a breath is not neutral or bland—it's *cooked air;* we live in a constant simmering. There is a furnace in our cells, and when we breathe we pass the world through our bodies, brew it lightly, and turn it loose again, gently altered for having known us.

A MAP OF SMELL

Breaths come in pairs, except at two times in our lives—the beginning and the end. At birth, we inhale for the first time; at death, we exhale for the last. In between, through all the lather of one's life, each breath passes air over our olfactory sites. Each day, we breathe about 23,040 times and move around 438 cubic feet of air. It takes us about five seconds to breathe—two seconds to inhale and three seconds to exhale—and, in that time, molecules of odor flood through our systems. Inhaling and exhaling, we smell odors. Smells coat us, swirl around us, enter our bodies, emanate from us. We live in a constant wash of them. Still, when we try to describe a smell, words fail us like the fabrications they are. Words are small shapes in the gorgeous chaos of the world. But they are shapes, they bring the world into focus, they corral ideas, they hone thoughts, they paint watercolors of perception. Truman Capote's *In Cold Blood* chronicles the mischief of two murderers who collaborated on a particularly nasty crime. A criminal psychologist, trying to explain the event, observed that neither one of them would have been capable of the crime separately, but together they formed a third person, someone who was able to kill. I think of metaphors as a more benign but equally potent example of what chemists call hypergolic. You can take two substances, put them together, and produce something powerfully different (table salt), sometimes even explosive (nitroglycerine). The charm of language is that, though it's human-made, it can on rare occasions capture emotions and sensations which aren't. But the physiological links between the smell and language centers of the brain are pitifully weak. Not so the links between the smell and the memory centers, a route that carries us nimbly across time and distance. Or the links between our other senses and language. When we see something, we can describe it in gushing detail, in a cascade of images. We can crawl along its surface like an ant, mapping each feature, feeling each texture, and describing it with visual adjectives like red, blue, bright, big, and so on. But who can map the features of a smell? When we use words such as smoky, sulfurous, floral, fruity, sweet, we are describing smells in terms of other things (smoke, sulfur, flowers, fruit, sugar). Smells are our dearest kin, but we cannot remember their names. Instead we tend to describe how they make us feel. Something smells "disgusting," "intoxicating," "sickening," "pleasurable," "delightful," "pulse-revving," "hypnotic," or "revolting."

My mother once told me about a drive she and my father took through the Indian River orange groves in Florida when the trees were thick with blossom and the air drenched with fragrance. It overwhelmed her with pleasure. "What does it smell like?" I asked. "Oh, it's delightful, an intoxicating delightful smell." "But what does that smell *smell* like?" I asked again. "Like oranges?" If so, I might buy her some eau de cologne, which has been made of neroli (attar of oranges), bergamot (from orange rind), and other minor ingredients since its creation in the eighteenth century, when it was the favorite of Madame du Barry. (Although the use of neroli itself as a perfume probably goes back to the days of the Sabines.) "Oh, no," she said with certainty, "not at all like oranges. It's a delightful smell. A wonderful smell." "Describe it," I begged. And she threw up her hands in despair.

Try it now. Describe the smell of your lover, your child, your parent. Or even one of the aromatic clichés most people, were they blindfolded, could recognize by smell alone: a shoe store, a bakery, a church, a butcher shop, a library. But can you describe the smell of your favorite chair, of your attic or your car? In *The Place in Flowers Where Pollen Rests*, novelist Paul West writes that "blood smells like dust." An arresting metaphor, one that relies on indirection, as metaphors of smell almost always do. Another engagingly subjective witness is novelist Witold Gombrowicz, who, in the first volume of his diary, recalls having breakfast at the Hermitage "with A. and his wife. . . . The food smells of, forgive me, a very luxurious water closet." I presume it was the fried kidneys for breakfast he didn't care for, even if they were expensive and high-class kidneys. For the cartography of smell, we need sensual mapmakers to sketch new words, each one precise as a landform or cardinal direction. There should be a word for the way the top of an infant's head smells, both talcumy and fresh, unpolluted by life and diet. Penguins smell starkly *penguin*, a smell so specific and unique that one succinct adjective should capture it. *Pinguid*, which means oily, won't do. *Penguinine* sounds like a mountain range. *Penguinlike* is the usual model, but it just clutters up the language and labels without describing. If there are words for all the pastels in a hue—the lavenders, mauves, fuchsias, plums, and lilacs—who will name the tones and tints of a smell? It's as if we were hypnotized en masse and told to selectively forget. It may be, too, that smells move us so profoundly, in part, because we cannot utter their names. In a world sayable and lush, where marvels offer themselves up readily for verbal dissection, smells are often right on the tip of our tongues—but no closer—and it gives them a kind of magical distance, a mystery, a power without a name, a sacredness.

OF VIOLETS AND NEURONS

Violets smell like burnt sugar cubes that have been dipped in lemon and velvet, I might offer, doing what we always do: defining one smell by another smell or another sense. In a famous letter, Napoleon told Josephine "not to bathe" during the two weeks that would pass before they met, so that he could enjoy all her natural aromas. But Napoleon and Josephine also adored violets. She often wore a violet-scented perfume, which was her trademark. When she died in 1814, Napoleon planted violets at her grave. Just before his exile on St. Helena, he made a pilgrimage to it, picked some of the violets, and entombed them in a locket, which he wore around his neck; they stayed there until the end of his life. The streets of nineteenth-century London were full of poor girls selling small bouquets of violets and lavender. In fact, Ralph Vaughan Williams's *London Symphony* includes an orchestral interpretation of the flower-girl's cry. Violets resist the perfumer's art and always have. It is possible to make a high-quality perfume from violets, but it's exceedingly difficult and expensive. Only the wealthiest people could afford it; but there have always been empresses, dandies, trend setters, and extravagants enough to keep perfumers busy. The thing about violets, which many people find cloying to the point of nausea, is that no response to them lasts long; as Shakespeare put it, they're:

> Forward, not permanent, sweet, not lasting,
> The perfume and suppliance of a minute.

Violets contain ionone, which short-circuits our sense of smell. The flower continues to exude its fragrance, but we lose the ability to smell it. Wait a minute or two, and its smell will blare again. Then it will fade again, and so on. How like Josephine, a woman of full-bodied if occasionally recondite sensuality, to choose as her trademark a scent that assaults the nose with a dam-burst of odor one second, and the next leaves the nose virginal, only to rampage yet again. No scent is more flirtatious. Appearing, disappearing, appearing, disappearing, it plays hide-and-seek with our senses, and there's no way to get too much of it. The violet so besotted the ancient Athenians that they chose it as their city's official flower and symbol. Victorian women liked to sweeten their breath with ca-chous, violet drops, especially if they'd been drinking. As I write this, I have been tasting a roll of "Choward's Violet" pastilles, "A deli-cious confection/Fragrance that refreshes," and the sweet, pun-gently musty ooze of violets has nearly swamped me. On the other hand, in the Amazon I brewed a pot of *casca preciosa*, a fragrant relative of the sassafras, whose steeped bark soon scented my face, my hair, my clothes, my room, and my psyche with hot violets of exquisite subtlety. If violets have thrilled, obsessed, repelled, and in other ways addled us for centuries, why is it so hard to describe them except indirectly? Do we smell indirectly? Not at all.

Smell is the most direct of all our senses. When I hold a violet to my nose and inhale, odor molecules float back into the nasal cavity behind the bridge of the nose, where they are absorbed by the mucosa containing receptor cells bearing microscopic hairs called cilia. Five million of these cells fire impulses to the brain's olfactory bulb or smell center. Such cells are unique to the nose. If you destroy a neuron in the brain, it's finished forever; it won't regrow. If you damage neurons in your eyes or ears, both organs will be irreparably damaged. But the neurons in the nose are replaced about every thirty days and, unlike any other neurons in the body, they stick right out and wave in the air current like anemones on a coral reef.

Found at the upper end of each nostril, the olfactory regions are yellow, richly moist, and full of fatty substances. We think of hered-ity as ordaining how tall one will be, the shape of the face, and the color of hair. Heredity also determines the shade of yellow of the olfactory area. The deeper the shade, the keener and more acute the sense of smell. Albinos have a poor sense of smell. Animals, which can smell with beatific grandeur, have dark-yellow olfactory regions; ours are light yellow. The fox's is reddish brown, the cat's an intense mustard brown. One scientist reports that dark-skinned men have darker olfactory regions and should therefore have more sensitive noses. When the olfactory bulb detects something—during eating, sex, an emotional encounter, a stroll through the park—it signals the cerebral cortex and sends a message straight into the limbic system, a mysterious, ancient, and intensely emotional section of our brain in which we feel, lust, and invent. Unlike the other senses, smell needs no interpreter. The effect is immediate and undiluted by language, thought, or translation. A smell can be overwhelmingly nostalgic because it triggers powerful images and emotions before we have time to edit them. What you see and hear may quickly fade into the compost heap of short-term memory, but, as Edwin T. Morris points out in *Fragrance,* "there is almost no short-term memory with odors." It's all long term. Furthermore, smells stimulate learning and retention. "When children were given olfactory information along with a word list," Morris noted, "the list was recalled much more easily and better retained in memory than when given without the olfactory cues." When we give perfume to someone, we give them liquid memory. Kipling was right: "Smells are surer than sights and sounds to make your heart-strings crack."

THE SHAPE OF SMELL

All smells fall into a few basic categories, almost like primary colors: minty (peppermint), floral (roses), ethereal (pears), musky (musk), resinous (camphor), foul (rotten eggs), and acrid (vinegar). This is why perfume manufacturers have had such success in concocting floral bouquets or just the right threshold of muskiness or fruitness. Natural substances are no longer required; perfumes can be made on the molecular level in laboratories. One of the first perfumes based on a completely synthetic smell (an aldehyde)* was Chanel No. 5,

---

*Aldehydes are a broad generic class of organic molecules, most of which are naturally occurring; rum and wine are flavored by wood aldehydes, which seep in from the keg.

which was created in 1922 and has remained a classic of sensual femininity. It has led to classic comments, too. When Marilyn Monroe was asked by a reporter what she wore to bed, she answered coyly, "Chanel No. 5." Its top note—the one you smell first—is the aldehyde, then your nose detects the middle note of jasmine, rose, lily of the valley, orris, and ylang-ylang, and finally the base note, which carries the perfume and makes it linger: vetiver, sandalwood, cedar, vanilla, amber, civet, and musk. Base notes are almost always of animal origin, ancient emissaries of smell that transport us across woodlands and savannas.

For centuries, people tormented and sometimes slaughtered animals to obtain four glandular secretions: ambergris (the oily fluid a sperm whale uses to protect its stomach from the sharp backbone of the cuttlefish and the sharp beak of the squid on which it feeds), castoreum (found in the abdominal sacs of Canadian and Russian beavers, and used by them to mark territories), civet (a honeylike secretion from the genital area of the nocturnal, carnivorous Ethiopian cat), and musk (a red, jellylike secretion from the gut of an East Asian deer). How did people first discover that the anal sacs of some animals held fragrance? Bestiality was common among shepherds in some of these regions, and it can't be ignored as one possibility. Because animal musk is so close to human testosterone, we can smell it in portions of as little as 0.00000000000032 of an ounce. Fortunately, chemists have now designed twenty synthetic musks, in part because the animals are endangered, and in part to ensure a consistency of odor difficult to achieve with natural substances. An obvious question is why secretions from the scent glands of deer, boar, cats, and other animals should arouse sexual desire in humans. The answer seems to be that they assume the same chemical shape as a steroid, and when we smell them we may respond as we would to human pheromones. In fact, in one experiment conducted at International Flavors and Fragrances, women who sniffed musk developed shorter menstrual cycles, ovulated more often, and found it easier to conceive. Does perfume matter—isn't it all packaging? Not necessarily. Can smells influence us biologically? Absolutely. Musk produces a hormonal change in the woman who smells it. As to why floral smells should excite us, well, flowers have a robust and energetic sex life: A flower's fragrance declares to all the world that it is fertile, available, and desirable, its sex organs oozing with nectar. Its smell reminds us in vestigial ways of fertility, vigor, life-force, all the optimism, expectancy, and passionate bloom of youth. We inhale its ardent aroma and, no matter what our ages, we feel young and nubile in a world aflame with desire.

Sunlight bleaches some of the smell from things, which anyone who has hung musty bedclothes on a clothesline in the sun will tell you. Even so, what remains might still smell stale and uninviting. We need only eight molecules of a substance to trigger an impulse in a nerve ending, but forty nerve endings must be aroused before we *smell* something. Not everything has a smell: only substances volatile enough to spray microscopic particles into the air. Many things we encounter each day—including stone, glass, steel, and ivory—don't evaporate when they stand at room temperature, so we don't smell them. If you heat cabbage, it becomes more volatile (some of its particles evaporate into the air) and it suddenly smells stronger. Weightlessness makes astronauts lose taste and smell in space. In the absence of gravity, molecules cannot be volatile, so few of them get into our noses deeply enough to register as odors. This is a problem for nutritionists designing space food. Much of the taste of food depends on its smell; some chemists have gone so far as to claim that wine is simply a tasteless liquid that is deeply fragrant. Drink wine with a head cold, and you'll taste water, they say. Before something can be tasted, it has to be dissolved in liquid (for example, hard candy has to melt in saliva); and before something can be smelled, it has to be airborne. We taste only four flavors: sweet, sour, salt, and bitter. That means that everything else we call "flavor" is really "odor." And many of the foods we think we can smell we can only taste. Sugar isn't volatile, so we don't smell it, even though we taste it intensely. If we have a mouthful of something delicious, which we want to savor and contemplate, we exhale; this drives the air in our mouths across our olfactory receptors, so we can smell it better.

But how does the brain manage to recognize and catalogue so

many smells? One theory of smell, J. E. Amoore's "stereochemical" theory, maps the connections between the geometric shapes of molecules and the odor sensations they produce. When a molecule of the right shape happens along, it fits into its neuron niche and then triggers a nerve impulse to the brain. Musky odors have disc-shaped molecules that fit into an elliptical, bowl-like site on the neuron. Pepperminty odors have a wedge-shaped molecule that fits into a V-shaped site. Camphoraceous odors have a spherical molecule that fits an elliptical site, but is smaller than that of musk. Ethereal odors have a rod-shaped molecule that fits a trough-shaped site. Floral odors have a disc-shaped molecule with a tail, which fits a bowl-and-trough site. Putrid odors have a negative charge that is attracted to a positively charged site. And pungent odors have a positive charge that fits a negatively charged site. Some odors fit a couple of sites at once and give a bouquet or blend effect. Amoore offered his theory in 1949, but it was also proposed in 60 B.C. by the wide-spirited poet Lucretius in his caravansary of knowledge and thought, *On the Nature of Things*. A lock-and-key metaphor seems increasingly to explain many facets of nature, as if the world were a drawing room with many locked doors. Or it may simply be that a lock and key is familiar imagery, one of the few ways in which human beings can make sense of the world around them (language and mathematics being two others). As Abram Maslow once said: If a man's only tool is a key, he will imagine every problem to be a lock.

Some smells are fabulous when they're diluted, truly repulsive when they're not. The fecal odor of straight civet would turn one's stomach, but in small doses it converts perfume into an aphrodisiac. Just a little of some smells—camphor, ether, oil of cloves for example—is too much, dulling the nose and making further smelling almost impossible. Some substances smell like other substances they seem remote from, in the nasal equivalent of referred pain (bitter almonds smell like cyanide; rotten eggs smell like sulfur). Many normal people have "blind spots," especially to some musks, and others can detect smells that are faint and fleeting. When we think of what's normal for human beings to sense, we tend to underimagine. One surprising thing about smell is the vast range of response one finds along the curve we call normal.

BUCKETS OF LIGHT

Much of life becomes background, but it is the province of art to throw buckets of light into the shadows and make life new again. Many writers have been gloriously attuned to smells: Proust's lime-flower tea and madeleines; Colette's flowers, which carried her back to childhood gardens and her mother, Sido; Virginia Woolf's parade of city smells; Joyce's memories of baby urine and oilcloth, holiness and sin; Kipling's rain-damp acacia, which reminded him of home, and the complex barracks smells of military life ("one whiff . . . is all Arabia"); Dostoevsky's "Petersburg stench"; Coleridge's note-books, in which he recalled that "a dunghill at a distance smells like musk, and a dead dog like elder flowers"; Flaubert's rhapsodic ac-counts of smelling his lover's slippers and mittens, which he kept in his desk drawer; Thoreau's moonlight walks through the fields when the tassels of corn smelled dry, the huckleberry bushes oozed musti-ness, and the berries of the wax myrtle smelled "like small confec-tionery"; Baudelaire's plunges into smell until his "soul soars upon perfume as the souls of other men soar upon music"; Milton's description of the odors God finds pleasing to His divine nostrils and those preferred by Satan, an ace sniffer-out of carrion ("Of carnage, prey innumerable . . . scent of living carcasses"); Robert Herrick's fetishistic and intimate sniffing of his sweetheart, whose "breast, lips, hands, thighs, legs . . . are all/ richly aromatical," indeed "All the spices of the East/ Are circumfused there"; Walt Whitman's praise of sweat's "aroma finer than prayer"; François Mauriac's *La Robe Prétexte,* which is adolescence remembered through its smells; Chaucer's "The Miller's Tale," where we find one of the first men-tions in literature of breath deodorants; Shakespeare's miraculously delicate flower similes (to the violet he says: "Sweet thief, whence didst thou steal the sweet, if not from my love's breath?"); Czeslaw Milosz's linen closet, "filled with the mute tumult of memories";

Joris-Karl Huysmans's obsession with nasal hallucinations, and the smell of liqueurs and women's sweat that fills his lush, almost unimaginably decadent, hedonistic novel, *A Rebours*. About one character, Huysmans explained that she was "an ill-balanced, nerve-ridden woman, who loved to have her nipples macerated in scents, but who really experienced a genuine and overmastering ecstasy when her head was tickled with a comb and she could, in the act of being caressed by a lover, breathe the smell of the chimney soot, of wet from a house building in rainy weather, or of dust of a summer storm."

The most scent-drenched poem of all time, "The Song of Solomon," avoids talk of body or even natural odors, and yet weaves a luscious love story around perfumes and unguents. In the story's arid lands, where water was rare, people perfumed themselves often and well, and this betrothed couple, whose marriage day approaches, in the meantime converse amorously in poetry, sweetly dueling with compliments lavish and ingenious. When he dines at her table he is "a bundle of myrrh" or "a cluster of camphire in the vineyards of En-ge-di," or muscular and sleek as a "young gazelle." To him, her robust virginity is a secret "garden . . . a spring shut up, a fountain sealed." Her lips "drop as the honeycomb: honey and milk are under thy tongue; and the smell of thy garments is like the smell of Lebanon." He tells her that on their wedding night he will enter her garden, and he catalogues all the fruits and spices he knows he'll find there: frankincense, myrrh, saffron, camphire, pomegranates, aloes, cinnamon, calamus, and other treasures. She will weave a fabric of love around him, and fill his senses until they brim with oceanic extravagance. So stirred is she by this loving tribute and so wild with desire that she replies yes, she will throw open the gates of her garden to him: "Awake, O north wind; and come, thou south; blow upon my garden, that the spices thereof may flow out. Let my beloved come into his garden, and eat his pleasant fruits."

In the macabre contemporary novel *Perfume*, by Patrick Süskind, the hero, who lives in Paris in the eighteenth century, is a man born without any personal scent whatsoever, although he develops prodigious powers of smell: "Soon he was no longer smelling mere wood, but kinds of wood: maple wood, oak wood, pinewood, elm wood, pearwood, old, young, rotting, moldering, mossy wood, down to single logs, chips, and splinters—and could clearly differentiate them as objects in a way that other people could not have done by sight." When he drinks a glass of milk each day, he can smell the mood of the cow it has come from; out walking, he can easily identify the origin of any smoke. His lack of human scent frightens people, who treat him badly, and this warps his personality. He ultimately creates personal odors for himself that other people aren't aware of per se, but which make him appear more normal, including such delicacies as "an odor of inconspicuousness, a mousey, workaday outfit of odors with the sour, cheesy smell of humankind still present." In time, he becomes a murderer-perfumer, who seeks to distill the fragrant essence from certain people as if they were flowers.

Many writers have written of how smells trigger flights of comprehensive remembrance. In *Swann's Way,* Proust, that great blazer of scent trails through the wilderness of luxury and memory, describes a momentary whirlwind in his day:

> I would turn to and fro between the prayer-desk and the stamped velvet armchairs, each one always draped in its crocheted antimacassar, while the fire, baking like a pie the appetizing smells with which the air of the room was thickly clotted, which the dewy and sunny freshness of the morning had already "raised" and started to "set," puffed them and glazed them and fluted them and swelled them into an invisible though not impalpable country cake, an immense puff-pastry, in which, barely waiting to savor the crustier, more delicate, more respectable, but also drier smells of the cupboard, the chest-of-drawers, and the patterned wall-paper I always returned with an unconfessed gluttony to bury myself in the nondescript, resinous, dull, indigestible, and fruity smell of the flowered quilt.

Throughout his adult life, Charles Dickens claimed that a mere whiff of the type of paste used to fasten labels to bottles would bring back with unbearable force all the anguish of his earliest years, when bankruptcy had driven his father to abandon him in a hellish ware-house where they made such bottles. In the tenth century, in Japan, a glitteringly talented court lady, Lady Murasaki Shikibu, wrote the first real novel, *The Tale of Genji*, a love story woven into a vast historical and social tapestry, the cast of which includes perfumer-alchemists, who concoct scents based on an individual's aura and destiny. One of the real tests of writers, especially poets, is how well they write about smells. If they can't describe the scent of sanctity in a church, can you trust them to describe the suburbs of the heart?

THE WINTER PALACE OF MONARCHS

We each have our own aromatic memories. One of my most vivid involves an odor that was as much vapor as scent. One Christmas, I traveled along the coast of California with the Los Angeles Mu-seum's Monarch Project, locating and tagging great numbers of overwintering monarch butterflies. They prefer to winter in eucalyp-tus groves, which are deeply fragrant. The first time I stepped into one, and every time thereafter, they filled me with sudden tender memories of mentholated rub and childhood colds. First we reached high into the trees, where the butterflies hung in fluttering gold garlands, and caught a group of them with telescoping nets. Then we sat on the ground, which was densely covered with the South African ice plant, a type of succulent, and one of the very few plants that can tolerate the heavy oils that drop from the trees. The oils kept crawling insects away, too, and, except for the occasional Pacific tree frog croaking like someone working the tumblers of a safe, or a foolish blue jay trying to feed on the butterflies (whose wings contain a digitalis-like poison), the sunlit forests were serene, other-worldly, and immense with quiet. Because of the eucalyptus vapor, I not only smelled the scent, I felt it in my nose and throat. The loudest noise was the occasional sound of a door creaking open, the sound of eucalyptus bark peeling off the trees and falling to the ground, where it would soon roll up like papyrus. Everywhere I looked, there seemed to be proclamations left by some ancient scribe. Yet, to my nose, it was Illinois in the 1950s. It was a school day; I was tucked in bed, safe and cosseted, feeling my mother massage my chest with Vicks VapoRub. That scent and memory brought an added serenity to the hours of sitting quietly in the forest and handling the exquisite butterflies, gentle creatures full of life and beauty who stalk nothing and live on nectar, like the gods of old. What made this recall doubly sweet was the way it became layered in my senses. Though at first tagging butterflies triggered memories of childhood, afterward the butterfly-tagging *itself* became a scent-triggerable memory, and, what's more, it replaced the original one: In Manhattan one day, I stopped at a flower-seller's on the street, as I always do when I travel, to choose a few flowers for the hotel room. Two tubs held branches of round, silver-dollar-shaped eucalyptus, the leaves of which were still fresh—bluish-green with a chalky surface; a few of them had broken, and released their thick, pungent vapor into the air. Despite the noise of Third Avenue traffic, the drilling of the City Works Department, the dust blowing up off the streets and the clotted gray of the sky, I was instantly transported to a particularly beautiful eucalyptus grove near Santa Barbara. A cloud of butterflies flew along a dried-up riverbed. I sat serenely on the ground, lifting yet another gold-and-black monarch butterfly from my net, carefully tagging it and tossing it back into the air, then watching for a moment to make sure it flew safely away with its new tag pasted like a tiny epaulet on one wing. The peace of that moment crested over me like a breaking wave and saturated my senses. A young Vietnamese man arranging his stock looked hard at me, and I realized that my eyes had suddenly teared. The whole episode could not have taken more than a few seconds, but the combined scent memories endowed eucalyptus with an almost savage power to move me. That afternoon, I went to one of my favorite shops, a boutique in the Village, where they will compound a bath oil for you, using a base of sweet almond oil, or make up shampoos or body lotions from other fragrant ingredients. Hanging from my bathtub's shower attachment is a blue net bag of the sort French-women use when they do their daily grocery shopping; I keep in it a wide variety of bath potions, and eucalyptus is one of the most calming. How is it possible that Dickens's chance encounter with a few molecules of glue, or mine with eucalyptus, can transport us back to an otherwise inaccessible world?

THE OCEANS INSIDE US

Driving through farm country at summer sunset provides a cavalcade of smells: manure, cut grass, honeysuckle, spearmint, wheat chaff, scallions, chicory, tar from the macadam road. Stumbling on new smells is one of the delights of travel. Early in our evolution we didn't travel for pleasure, only for food, and smell was essential. Many forms of sea life must sit and wait for food to brush up against them or stray within their tentacled grasp. But, guided by smell, we became nomads who could go out and search for food, hunt it, even choose what we had a hankering for. In our early, fishier version of humankind, we also used smell to find a mate or detect the arrival of a barracuda. And it was an invaluable tester, allowing us to prevent something poisonous from entering our mouths and the delicate, closed system of our bodies. Smell was the first of our senses, and it was so successful that in time the small lump of olfactory tissue atop the nerve cord grew into a brain. Our cerebral hemispheres were originally buds from the olfactory stalks. We *think* because we *smelled.*

Our sense of smell, like so many of our other body functions, is a throwback to that time, early in evolution, when we thrived in the oceans. An odor must first dissolve into a watery solution our mucous membranes can absorb before we can smell it. Scuba-diving in the Bahamas some years ago, I became aware of two things for the first time: that we carry the ocean within us; that our veins mirror the tides. As a human woman, with ovaries where eggs lie like roe, entering the smooth, undulating womb of the ocean from which our ancestors evolved millennia ago, I was so moved my eyes teared underwater, and I mixed my saltiness with the ocean's. Distracted by such thoughts, I looked around to find my position vis-à-vis the boat, and couldn't. But it didn't matter: Home was everywhere.

That moment of mysticism left my sinuses full, and made surfacing painful until I removed my mask, blew my nose in a strange two-stage snite, and settled down emotionally. But I've never forgotten that sense of belonging. Our blood is mainly salt water, we still require a saline solution (salt water) to wash our eyes or put in contact lenses, and through the ages women's vaginas have been described as smelling "fishy." In fact, Sandor Ferenczi, a disciple of Freud's, went so far as to declare, in *Thalassa: A Theory of Genitality*, that men only make love to women because women's wombs smell of herring brine, and men are trying to get back to the primordial ocean—surely one of the more remarkable theories on the subject. He didn't offer an explanation for why women have intercourse with men. One researcher claims that this "fishiness" is due not to anything intrinsic to the vagina, but rather to poor hygiene after intercourse, or vaginitis, or stale sperm. "If you deposit semen in the vagina and leave it there, it comes out smelling fishy," he argues. This has a certain etymological persuasiveness to it, if we remember that in many European languages the slang names for prostitutes are variations on the Indo-European root *pu*, to decay or rot. In French, *putain*; to the Irish, *old put*; in Italian *putta*; *puta* in both Spanish and Portuguese. Cognate words are putrid, pus, suppurate, and putorius (referring to the skunk family). *Skunk* derives from the Algonquin Indian word for polecat; and during the sixteenth and seventeenth centuries in England polecat was a derogatory term for prostitute. Not only do we owe our sense of smell and taste to the ocean, but we smell and taste *of* the ocean.

NOTIONS AND NATIONS OF SWEAT

In general, humans have a strong body odor, and anthropologist Dr. Louis S. B. Leakey thinks our ancestors may have had an even stronger odor, one that predatory animals found foul enough to avoid. Not long ago, I spent some time in Texas, studying bats. I placed a large Indonesian flying fox in my hair, to see if it would get entangled, as the old wives' tales warned. Not only did it not tangle, it began to cough gently from the mingling smells of my soap,

cologne, saltiness, oils, and other human odors. When I put it back
in its cage, it cleaned itself like a cat for many minutes, clearly feeling
soiled by the human contact. Many plants—like rosemary or sage—
have evolved pungent odors to repel predators; why not animals?
Nature rarely wastes a winning strategy. Of course, some humans
have much stronger odors than others. Folk wisdom says that bru-
nettes "smell different" from redheads, who smell different from
blondes. There's been so much anecdotal evidence about different
races having distinctive odors—because of diets, habits, hairiness or
lack of it—that such claims are difficult to discount, even though the
topic scares most scientists, who are understandably concerned
about being called racist.* There hasn't been a great deal of research
into national and racial odors. In any case, one culture doesn't
"smell" better or worse than another, just different, but that may be
why the word "stinking" so often appears as an adjective in streams
of racial abuse. Asiatics don't have as many apocrine glands at the
base of hair follicles as occidentals do, and as a result they often find
Europeans ripe-smelling. A strong body odor among Japanese men
is so rare that at one time it could disqualify them from military
service. This is also why there is so much scenting of the room and
air in Asian life, and much less scenting of the body. Pungent odors
are absorbed by fats: If you put an onion or cantaloupe in the
refrigerator with an open tub of butter, the butter will absorb the
odor. Hair also contains fat, which is why it leaves grease stains on
pillows and antimacassars. It absorbs smells, too, like smoke or cologne. The hairiness of Caucasians and Blacks makes them very
sweaty compared to Asians, but colognes simmer in their oil and
warmth like votive candles.

Body odor comes from the apocrine glands, which are small when
we're born and develop substantially during puberty; there are many

---

*The authors of a paper in *Science* a few years ago discovered that some black men appear
to have larger penises than white men—that is, the penis appears larger when in repose,
because the gene that carries sickle-cell anemia tends to make the penis semi-erect when it's
flaccid. I was told that the authors of the study had hesitated for some time before publishing
their findings, and then did so anxiously and with misgivings.

of them scattered around our armpits, face, chest, genitals, and anus. Some researchers conclude that a large part of our joy in kissing is really a joy in smelling and caressing each other's face, where one's personal scent glows. Among far-flung tribes in a number of countries—Borneo, on the Gambia River in West Africa, in Burma, in Siberia, in India—the word for "kiss" means "smell"; a kiss is really a prolonged smelling of one's beloved, relative, or friend. Members of a tribe in New Guinea say good-bye by putting a hand in each other's armpit, withdrawing it and stroking it over themselves, thus becoming coated with the friend's scent; other cultures sniff each other or rub noses in greeting.

THE PERSONALITY OF SMELL

Meat eaters smell different from vegetarians, children smell different from adults, smokers smell different from nonsmokers; other individuals smell different because of hereditary factors, health, occupation, diet, medication, emotional state, even mood. As Roy Bedichek observes in *The Sense of Smell:* "The body odor of his prey excites the predator so that his mouth waters and every fiber of his being becomes taut and every sense alerted. At the same time in the nostrils of the prey, fear and hate become associated with the body odor of the predator.* Thus on low levels of animal life, a specific odor evolves along with and becomes identified with a specific mood." Each person has an odor as individual as a fingerprint. A dog can identify it easily and recognize its owner even if he or she is one of a pair of identical twins. Helen Keller swore that by simply smelling people she could decipher "the work they are engaged in. The odors of the wood, iron, paint, and drugs cling to the garments of those who work in them. . . . When a person passes quickly from one place to another, I get a scent impression of where he has been—the kitchen, the garden, or the sickroom."

---

*Novelists have written about the smell of fear, and researchers working with rats have found that stressed rats give off a special odor. Other unstressed rats detect the odor and have a physical, analgesic response, so that they will be prepared for pain.

For those of exquisite sensuality, there is nothing headier than the musky smell of a loved one moist with sweat. But natural body odors don't strike most of us as particularly enticing. In the Elizabethan Age, lovers exchanged "love apples"—a woman would keep a peeled apple in her armpit until it was saturated with her sweat, and then give it to her sweetheart to inhale. Now we have whole industries devoted to removing our natural odors and replacing them with artificial ones. Why do we prefer our breath to smell of peppermint instead of rotting bacteria, our "natural" smell? True, a foul smell might signal disease: We might not be attracted to someone giving off an unhealthy odor, and an excess of rotting bacteria could persuade us we are chatting with, say, a cholera victim, someone who could infect us. But mainly we value one scent over another thanks to Madison Avenue's brashness and our gullibility. Aromatic paranoia pays well. In creative greed, they've frightened us into thinking that we're "offensive" and require lotions and potions to mask our natural odors.

Just what do we mean by a bad smell? And what is the worst smell in the world? The answers depend on culture, age, and personal taste. Westerners find fecal smells repulsive, but the Masai like to dress their hair with cow dung, which gives it an orangey-brown glow and a powerful odor. Children like most smells until they're old enough to be taught differently. When naturalist and zookeeper Gerald Durrell wanted to catch some fruit bats for his zoo on the Isle of Jersey, he went to the island of Rodriguez, east of Madagascar, and baited his net with what he called "jackfruit," a big, brown durianlike hedgehog of a fruit, whose white pulp reeked "like a cross between an open grave and a sewer," a regular "charnel house." That sounds pretty bad to me, and so, just to see if he's right, I've put "Rodriguez in jackfruit season" on the long list of sensory destinations I'd like to get to one day.

Though ancient and uncontrollably natural, a fart is generally considered to be repellant, discourteous, and even the smell of the devil. *The Merck Manual,* in an uncharacteristically entertaining chapter on "Functional Bowel Disease," subheading "Gas," describes the possible origins, treatments of, and miscellaneous symptoms and signs of gas, along with this observation:

> Among those who are flatulent, the quantity and frequency of gas passage can reach astounding proportions. One careful study noted a patient with daily flatus frequency as high as 141, including 70 passages in one 4-h period. This symptom, which can cause great psychosocial distress, has been unofficially and humorously described according to its salient characteristics: (1) the "slider" (crowded elevator type), which is released slowly and noiselessly, sometimes with devastating effect; (2) the open sphincter, or "pooh" type, which is said to be of higher temperature and more aromatic; and (3) the staccato or drum-beat type, pleasantly passed in privacy.
>
> While questions of air pollution and degradation of air quality have been raised, no adequate studies have been performed. However, no hazard is likely to those working near open flames, and youngsters have even been known to make a game of expelling gas over a match-flame. Rarely, this usually distressing symptom has been turned to advantage, as with a Frenchman referred to as "Le Pétomane," who became affluent as an effluent performer on the Moulin Rouge stage.

In his fascinating history of stench, perfume, and society in France, *The Foul and the Fragrant,* Alain Corbin describes the open sewers of Paris at the time of the revolution, and points out how strong a role scent has also played in fumigation throughout history. There are various forms of fumigation—fumigation for health reasons (especially during plagues); insect fumigation; and even religious and moral fumigation. The floors of medieval castles were strewn with rushes, lavender, and thyme, which were thought to prevent typhus. Perfumes were often used for magical and alchemical purposes, too, promising an enchantment. If the promises of today's perfume ads seem extravagant, consider those made in the sixteenth century. In *Les secrets de Maistre Alexys le Piedmontois,* a book on cosmetics, the author promises that his toilet water will make women not just attractive for an evening but beautiful "forever." "Forever" is pretty serious advertising, and probably should tip off a potential consumer to read the fine print. Here is the ghoulish recipe: "Take a young raven from its nest, feed it on hard-boiled eggs for forty days, kill it, then distill it with myrtle leaves, talcum powder, and almond oil." Splendid. Except for the stench, and an overwhelming desire to quote Poe, you'll surely be a ravenous beauty perching on the eaves of forever.

PHEROMONES

Pheromones are the pack animals of desire (from Greek, *pherein,* to carry, and *horman,* excite). Animals, like us, not only have distinctive odors, they also have powerfully effective pheromones, which trigger other animals into ovulation and courtship, or establish hierarchies of influence and power. They scent-mark, sometimes in ingenious ways. Voles and bush babies spray the soles of their feet with urine and brand the earth with it as they patrol their territories. Antelopes mark trees using scent glands on their faces. Cats have scent glands on their cheeks, and can often be seen "cheeking" someone or a favorite table leg. When you pet a cat, she will, if she likes you, lick herself to taste your scent. And then she'll probably choose your favorite armchair to claw and curl up in, not just because of its cushions but because your scent is on it. The polecat, as well as the badger, drags its anus along the ground to mark it. Jane Goodall, in *The Innocent Killers,* reports that male and female wild dogs scent-mark one after the other on exactly the same blades of grass, to inform all interested parties that they are a pair. When my friend takes her German shepherd Jackie out for a walk, Jackie sniffs at curb, rock, and tree, and soon senses what dog has been there, its age, sex, mood, health, when it last passed by. For Jackie, it's like reading the gossip column of the morning newspaper. The lane reveals its invisible trails to her nose as it doesn't to her owner. She will add her scent to the quilt of scents on a tuft of grass, and the next dog that comes along will read, in the aromatic hieroglyphics of the neighborhood, *Jackie, 5:00 P.M., young female, on hormone therapy because of a bladder ailment, well fed, cheerful, seeks a friend.*

Sometimes messages can't be merely immediate; they need to last over time, and yet be a constant signal, like a lighthouse guiding animals through the breakwaters of their uncertainty. Most smells will glow for a while, where a wink may vanish before it's seen, a flexed muscle imply too many things, a voice startle or threaten. For an animal who is prey, the odor of its hunter will warn it; for the hunter, the odor of its prey will lure it. Of course, some animals exude an odor as a form of defense. Spotted skunks do a handstand and squirt would-be attackers with a horrible stench. Among insects, odor is all forms of communication: a guidebook to nesting or egg-laying spots, a rallying cry, a trumpet flourish announcing royalty, an alarm warning of ambush, a map home. In the rain forest, one can see long, ropy caravans of ants, marching single file along trails of scent that have been laid down for them by scouts. They may seem to be scrambling around in a blind fury of industriousness, but they are always in touch with one another, always gabbing about something meaningful to their lives. A male butterfly of the Danaidae family travels from flower to flower, mixing a cocktail of scents in a pocket on each hind leg until he has the perfect perfume to attract a female.* Birds sing to announce their presence in the world, mark their territories, impress a mate, boast of their status—ultimately, much of it has to do with sex and mating. Mammals prefer to use odors when they can, spinning scent songs as complex and unique as bird songs, which also travel on the air. Baby kangaroos, puppies, and many other mammals are born blind and must find their way to the nipple by smell. A mother fur seal will go out fishing, return to a beach swarming with pups, and recognize her own partly by smell. A mother bat, entering a nursery cave where millions of mother and baby bats cling to the wall or wing through the air, can find her young by calling to it and smelling a path toward it. When I was on a cattle ranch in New Mexico, I often saw a calf with the skin of another calf tied around its back, nursing happily. A cow recognizes her calf by smell, which triggers her mothering instincts,

---

*Butterflies often give off an aroma to attract a mate, and may smell like roses, sweetbriar, heliotrope, and other flowers.

so whenever there was a stillborn, the rancher would skin the dead calf and give its scent to an orphan.

Animals would not be able to live long without pheromones because they couldn't mark their territories or choose receptive, fertile mates. But are there human pheromones? And can they be bottled? Some trendy women in Manhattan are wearing a perfume called Pheromone, priced at three hundred dollars an ounce. Expensive perhaps, but what price aphrodisia? Based on findings about the sexual attractants animals give off, the perfume promises, by implication, to make a woman smell provocative and turn stalwart men into slaves of desire: love zombies. The odd thing about the claims of this perfume is that its manufacturer has not specified *which* pheromones are in it. Human pheromones have not yet been identified by researchers, whereas, say, boar pheromones have. The vision of a generation of young women walking the streets wearing boar pheromones is strange, even for Manhattan. Let me propose a naughty recipe: Turn loose a herd of sows on Park Avenue. Mix well with crowds of women wearing Pheromone eau de cologne. Dial 911 for emergency.

If we haven't yet pinpointed human pheromones, surely we can just use our secretions the way animals do, bottle our effluvia at different times of the month. Avery Gilbert, a biophysiologist, doesn't think so. It's more like psychology in a vial. He told *Gentleman's Quarterly* that "If you had a bottle full of fluids generated by the female genital glands during copulation, and you put it on a guy's desk, and if he even recognized the odor, he'd be *embarrassed*. Because it's out of context, and that's what makes the difference. If male consumers actually believe a claim that this component will get women hot, then they're naïve. I don't think there is a chemical that will do that. But it may not be important what particular odor men are broadcasting; it's the signal of availability, the perception of self-confidence. Those claims are implied and probably work. And that's probably the basic reason people wear the stuff."

One of Gilbert's colleagues, George Preti, staged an experiment in which ten women had the sweat of other women applied under their noses at regular intervals. It took three months for the women to begin menstruating at the same time as the women whose sweat they were smelling. A control group, daubed with alcohol instead of sweat, didn't change their cycles at all. Clearly, a pheromone in sweat affects menstrual synchrony, which is why women in dorms or close girlfriends so often menstruate at the same time, a phenomenon known as the McClintock Effect (after Martha McClintock, the psychologist who first observed it). There appear to be other effects. When a man gets involved with a woman for any length of time, his facial hair starts to grow faster than it did before. Women who are cloistered away from men (in a boarding school, say), enter puberty later than women who are around men. Mothers recognize the odor of their newborn children, and vice versa, so some doctors are experimenting with giving children bursts of their mother's odor, along with the anesthetic, during operations. Babies can smell their mother entering a room, even if they can't see her. In J. M. Barrie's *Peter Pan*, children can even "smell danger" while they sleep. Mothers of school-age children can pick out T-shirts worn by their own child. This is not true for fathers, who do not recognize the smell of their infants, but men can determine whether a T-shirt has been worn by a male or a female. Pheromones do affect people. But how much? Do pheromones trigger vigorous responses in us as they do in moths or beavers, or do they figure in the cascade of our sensory awareness no more significantly than ordinary visual or hearing cues? If I see a handsome man with beautiful blue eyes, am I having a "visualmone," as one researcher called it dismissively, or is it just that blue eyes excite me because they register as attractive in the culture, time, and context of my life? Blue eyes, "baby blues," remind us a little of Caucasian newborns, and fill us with protectiveness. But in some African cultures they would be thought ghoulish, icy, and unattractive.

Science fiction has often frightened us with humans as automatons, driven by unknown forces, their minds a sort of dial tone. Suppose pheromones at times secretly cancel our powers of choice and decision? The idea alarms. We don't like to lose control, except on purpose—during sex or partying or religious mysticism or doing drugs—and then only because we believe we're just fractionally more in control than we're not, or at least that such control will return to us quickly. Evolution is complex and at times amusing, so much of an adventure that few of its whims or obbligatos frighten me. Our apparent need for violence does, but not the possibility that we might be having elaborate, if subtle, conversations with one another through pheromones. Free will may not be entirely free, but it certainly is willful, and yet it seems as if there is a good deal of stretch in it. Such masterly ad-libbers as human beings know how to revise on almost any theme. If there's one thing at which we really excel, it is at pushing limits, inventing strategies, finding ways to sidestep the rudest truths, grabbing life by the lapels and shaking it soundly. Granted, it tends to shake back, but that never stops us.

NOSES

When we crawled or flopped out of the ocean onto the land and its trees, the sense of smell lost a little of its urgency. Later, we stood upright and began to look around, and to climb, and what a world we discovered spread out before us like a field of Texas bluebonnets! We could see for miles in all directions. Enemies became visible, food became visible, mates became visible, trails became visible. The shadow of a distant lion slinking through the grass was a more useful sign than any smell. Vision and hearing became more important for survival. Monkeys don't smell things as well as dogs do. Most birds don't have very sophisticated noses, although there are some exceptions—New World vultures locate carrion by smell, and seabirds often navigate by smell. But the animals with the keenest sense of smell tend to walk on all fours, their heads hanging close to the ground, where the damp, heavy, fragrant molecules of odor lie. This includes snakes and insects, too, along with elephants (whose trunks hang low), and most quadrupeds. Pigs can smell truffles under six inches of soil. Squirrels find nuts they buried months earlier. Bloodhounds can smell a man's scent in a room he left hours before, and then track the few molecules that seep through the soles of his shoes and land on the ground when he walks, over uneven terrain, even on stormy nights. Fish need olfactory abilities: Salmon can smell the distant waters of their birth, toward which they must swim to spawn. A male butterfly can home in on the scent of a female that is miles away. Pity us, the long, tall, upright ones, whose sense of smell has weakened over time. When we are told that a human has five million olfactory cells, it seems like a lot. But a sheepdog, which has 220 million, can smell forty-four times better than we can. What does it smell? What are we missing? Just imagine the stereophonic world of aromas we must pass through, like sleepwalkers without headphones. Still, we do have a remarkably detailed sense of smell, given how small our organs of smell really are. Because our noses jut out from our faces, odors have quite a distance to travel inside them before we're aware of what the nose has probed. That's why we wrinkle up our noses and sniff—to move the molecules of smell closer to the olfactory receptors hidden awkwardly in the backmost recesses of the nose.

SNEEZING

Few pleasures are as robust as the simple country pleasure of sneezing. The whole body ripples in orgasmic delight. But only humans sneeze with their mouths open. Dogs, cats, horses, and most other animals just sneeze straight down their noses, with the air bending a little at the neck. But humans huff and tremble in an anticipatory itch, draw in a big gobful of air, contract the ribs and stomach like a bellows, and violently shoot air into the nose, where it stops short, blasts the general area, and sometimes sprays messily out of the nose and mouth all at once. This wouldn't matter too much if our lungs blew air out gently during a sneeze. But researchers at the University of Rochester have found that a sneeze expels the air at eighty-five percent the speed of sound, fast enough to scour bacteria and other detritus from the body, the sneeze's goal. Human noses have a hairpin turn way at the back of the nasal passages, which makes the whole process of breathing more taxing, and inhaling odor molecules more difficult. There is no direct path for the air to follow in a sneeze. We have to open our mouths. If we sneeze closed-mouthed, the air thunders around the cavities and passages in our heads, looking for a way out, and can hurt our ears. There are many theories about why our noses are so poorly designed; in the last analysis, it probably has to do with the evolution of our biggish brains and the cramped space in our skulls, and to permit stereo vision. Bedichek suggests that the design didn't become awkward until we "swarmed into those congested areas we call 'cities.' Here the nose has had forced upon it suddenly a function it was never intended to perform, namely, screening out dust and grit while at the same time being subjected to intolerable odors of municipal filth, and finally to fumes from the vast chemical laboratory the modern city has become." The seventeenth-century poet Abraham Cowley states the point as a rhetorical question:

> Who that has reason, and his smell,
> Would not among roses and jasmine dwell,
> Rather than all his spirits choke
> With exhalations of dirt and smoke?

A tickle is all it takes. Or the sun. Some people, like me, inherit a genetic oddity that causes them to sneeze when confronted by bright light. I'm afraid this syndrome has been given the overly cute acronym of ACHOO (*a*utosomol *d*ominant *c*ompelling *h*elio-*o*phthalmic *o*utburst). If I feel a sneeze hovering, all I have to do is look at the sun to bring on the explosion, a light apocalypse.

SMELL AS CAMOUFLAGE

Though it's April, we've had snow in Ithaca for weeks, or so my neighbor tells me—I was in Manhattan, a maritime climate. Now I find that small mute deer prints lead right up to the door and the huge windows, dart across the frozen pool sparkling with rime, then meander through drifts to twin apple trees and ice-claggy fruit. So they have learned how to walk on water, browse the fragrant marvels tucked beneath the surface of the world, even how best to come and go in a season oblique with bullets and ice. Did they search for me, where I used to pause, reflecting in the glass? What if, later this spring, the frozen pool plays tricks and sags beneath their hooves, then folds up over them, and I do not hear their underwater screams? What if, like the snow, I have drifted too far? Craving the dialect of cities, I forgot the way deer steal into the yard with their big hearts and fragile dreams. I wasn't here to follow their gaunt, level eyes, or the staggering poetry of their hooves.

Often, I see them browsing in the yard, but when I slip outside for a closer look they smell my strong human scent, amble down to the fence, and leap back into their pandemonium of green. This summer I intend to disguise myself as a conifer or a mushroom. A recent issue of *Field and Stream* tells me how: To fool deer and rabbits, take something without much tannin (yellow birch, pine, mushrooms, hemlock, wintergreen, or some aromatic conifer, for example) and dry it for a week or two. Chop it up, then fill a jar half full of it. Add 100-proof vodka. Filter through a Melitta filter. Put in an atomizer. Apply liberally to bury your human smell. Let your thoughts mushroom.

ROSES

I am holding a lavender rose called "Angel Face," one of the twenty-five rosebushes planted around my house. For the first few years, the deer that frequent my yard would steal in at dawn and eat all the buds and succulent new growth. Once they ate the bushes right down to the dirt, leaving only small knobs that looked like the velvet of incipient antlers. I am used to embezzlers in the garden. The first summer of the grape arbor, I watched two vines evolve from flowers to succulent purple fruits, sense-luscious and nearly bursting with fragrance. Each day, I watched them, waiting until the perfect moment of ripeness, imagining how it would be to roll the grapes around on my tongue, fresh, sweet, and quenching. One day the grapes' purple sheen changed to a taut, robust iridescence, and I knew the next morning would be the earliest day to pick. Such knowledge was not reserved for me alone. When I awoke, I found every single grape sucked dry, the skins littering the ground like tiny purple prepuces. This scene, left by raccoons, has repeated every autumn ever since, despite cages, cowbells, barbed wire, and other "deterrents," and frankly I've given up on grapes and raccoons. The roses pose a trickier problem.

I love the deer as well as the roses, so I decided to use smell as a weapon—after all, plants do it—and sprinkled a mixture of tobacco and naptha around the rosebushes. It worked, but made the air raunchy and caustic. Unless you crave the smell of baseball players at winter camp, their mouths full of chewing mess, their pockets full of mothballs. This year I have another plan: lavender. Deer hate its strong nose-scrubbing smell; I've ordered dozens of bushes to plant around the roses and day lilies, hoping they'll make an olfactory fence when the deer come calling. Still, we'll divide the spoils. I have left them the luxuriant raspberry bushes, which I no longer try to harvest, and the twin apple trees. The raccoons get the grape arbor, the rabbits get the wild strawberries. But the roses are sacrosanct, because they so drench my senses with exquisite smells. The most expensive perfume in the world, and one of the enduring classics, Joy, is a blend of two floral notes: jasmine and lots of rose.

Roses have tantalized, seduced, and intoxicated people more than any other flower. They've captivated homeowners, swains, flower addicts, and sensuists since the ancients. In Damascus and Persia, people used to bury jars of unopened rosebuds in the garden, and dig them up on special occasions to use in cooking—the flowers would open dramatically on the plates. In Jean Cocteau's film version of the fairy tale *Beauty and the Beast,* all the mischief and magic begins when a man picks a rose for his daughter, her sole desire among a sea chest of riches. Long ago, Europeans raised a tough mongrel rose that was loud, obvious, and very hardy, and whose fragrance could embalm a statue. But, in the 1800s, they began importing elegant Chinese tea roses, which smelled like fresh tea leaves when crushed, and also frost-delicate, ever-blooming Chinese hybrids with bright yellow to red flowers. Breeding the hybrid Chinas with the European roses as carefully as racehorses, they produced subtle and sophisticated offspring roses, charmed into a seemingly endless array of colors, shapes, and scents. They called them "hybrid tea roses." Since then, over twenty thousand varieties have been bred, and at one time the rose's fragrance was nearly lost through overbreeding. Fragrance seems to be a recessive trait in roses, and two deeply fragrant parents may produce a petal-perfect but smell-less offspring. Now the trend is toward perfumed roses, thank heavens. The most popular hybrid tea in the world is "Peace," a stunning multicolored pastel with sunset hues that shriek at noon, grow muted at sunset, and record all the other phantoms of light during the day. Its egg-shaped buds open into large, pale-yellow ruffles with translucent tips that are often flushed with pink. And it smells like sugared leather dipped in honey. Of all my roses, "Peace" seems to have an almost human complexion and human moods, depending on the moisture and light of each day. An experimental rose, it was named on May 2, 1945 (the day Berlin fell), at the Pacific Rose Society in Pasadena, because "this greatest new rose of our time should be named for the world's greatest desire—Peace." Many presidents have had roses named after them (Lincoln's is blood red, John Kennedy's pure white), and there are wittily named roses to honor movie stars or celebrities (Dolly Parton's is flamboyantly pungent, with knockout-sized blossoms). Though roses symbolize beauty and love, their colors, textures, shapes, and smells are difficult to describe. "Sutter's Gold," one of my favorite hybrid tea roses, produces a flat ruffled flower of yellow petals tinged in apricot, fuchsia, and pink, with a fragrance like sweet wet feathers. The floribundas, thoroughly modern roses, cascade with flowers all summer long. "The Fairy" has hardly any scent, but is a constant explosion of dainty pink flowers from spring until winter, despite light snowfalls. Roses were already considered ancient when the Greek botanist Theophrastus wrote about "the hundred-petaled rose" in 270 B.C. Fossilized wild roses have been dated as far back as forty million years ago. The Egyptian rose was what we now call the cabbage rose, renowned for its many petals. When Cleopatra welcomed Mark Antony to her bedroom, the floor was covered in a foot and a half of such petals. Did they use the floor, and make love in a swamp of soft, fragrant, shimmying petals? Or did they use the bed, as if they were on a raft floating in a scented ocean?

Cleopatra knew her guest. Few people have been as obsessed with roses as the ancient Romans. Roses were strewn at public ceremonies and banquets; rose water bubbled through the emperor's fountains and the public baths surged with it; in the public amphitheaters, crowds sat under sun awnings steeped in rose perfume; rose petals were used as pillow stuffings; people wore garlands of roses in their hair; they ate rose pudding; their medicines, love potions, and aphrodisiacs all contained roses. No bacchanalia, the Romans' official orgy, was complete without an excess of roses. They created a holiday, Rosalia, to formally consummate their passion for the flower. At one banquet, Nero had silver pipes installed under each plate, so that guests could be spritzed with scent between courses. They could admire a ceiling painted to resemble the celestial heavens, which would open up and shower them in a continuous rain of perfume and flowers. At another, he spent the equivalent of $160,000 just on roses—and one of his guests smothered to death under a shower of rose petals.

Islamic cultures found the rose a more spiritual symbol, one that, according to the thirteenth-century mystic Yunus Emre, is supposed to sigh "Allah, Allah!" each time one smells it. Mohammed, a great devotee of perfume, once said that the excellence of the extract of violets above all other flowers was like his own excellence above all other men. Nonetheless, it was rose water that went into the mortar for his temples. Roses mix unusually well with water, making fine sherbets and pastries, so the flower has become a delicate staple in Islamic cooking as well as being much used to scent apparel. Hospitality still demands that a guest in an Islamic household be sprinkled with rose water as soon as she or he arrives.

Rosaries originally consisted of 165 dried, carefully rolled-up rose petals (some of which were darkened with lampblack as a preservative) and the rose was the symbol of the Virgin Mary. When the crusaders returned to Europe, their senses sated by the exotic indulgences they discovered among the infidels, they brought attar of roses with them, along with sandalwood, pomander balls, and other rich spices and scents, plus a memory of harem women, sensual and languorous, who awaited a man's pleasure. The scented oils the knights returned with became instantly fashionable, suggesting all the wicked pleasures of the East, as seductive and irresistible as they were forbidden. Pleasures as sense-bludgeoning as a rose.

THE FALLEN ANGEL

Smells spur memories, but they also rouse our dozy senses, pamper and indulge us, help define our self-image, stir the cauldron of our seductiveness, warn us of danger, lead us into temptation, fan our religious fervor, accompany us to heaven, wed us to fashion, steep us in luxury. Yet, over time, smell has become the least necessary of our senses, "the fallen angel," as Helen Keller dramatically calls it. Some researchers believe that we do indeed perceive, through smell, much of the same information lower animals do. In a room full of businesspeople, one would get information about which individuals were important, which were confident, which were sexually receptive, which in conflict, all through smell. The difference is that we don't have a trigger response. We're aware of smell, but we don't automatically react in certain ways because of it, as most animals would.

One morning I took a train to Philadelphia to visit the Monell Chemical Senses Center near the campus of Drexel University. Laid out like a vertical neighborhood, Monell's building houses hundreds of researchers who study the chemistry, psychology, healing properties, and odd characteristics of smell. Many of the news-making pheromone studies have taken place at Monell, or at similar institutions. In one experiment, rooms full of housewives were paid to sniff anonymous underarms; in another study, funded by a feminine hygiene spray manufacturer, the scene was even more bizarre. Among

Monell's concerns: how we recognize smells; what happens when someone loses their sense of smell; how smell varies as one grows older; ingenious ways to control wildlife pests through smell; the way body odors can be used to help diagnose diseases (the sweat of schizophrenics smells different from that of normal people, for example); how body scents influence our social and sexual behavior. Monell researchers have discovered, in one of the most fascinating smell experiments of our time, that mice can discriminate genetic differences among potential mates by smell alone; they read the details of other animals' immune systems. If you want to create the strongest offspring, it's best to mate with someone whose strengths are different from yours, so that you can create the maximum defenses against any intruder, bacteria, viruses, and so on. And the best way to do that is to produce an omnicompetent immune system. Nature thrives in mongrels. *Mix well* is life's motto. Monell scientists have been able to raise special mice that differ from one another in only a single gene, and observe their mating preferences. They all chose mates whose immune systems would combine with theirs to produce the hardiest litters. Furthermore, they did not base their choices on their perception of their own smell, but on the remembered smell of their parents. None of this was reasoned, of course; the mice just mated according to their drive, unaware of the subliminal fiats.

Can it be possible that human beings do this, too, without realizing it? We don't require smell to mark territories, establish hierarchies, recognize individuals or, especially, know when a female is in heat. And yet one look at the obsessive use of perfume and its psychological effect on us makes it clear that smell is an old war-horse of evolution we groom and feed and just can't let go of. We don't need it to survive, but we crave it beyond all reason, maybe, in part, out of a nostalgia for a time when we were creatureal, a deeply connected part of Nature. As evolution has phased out our sense of smell, chemists have labored to restore it. Nor is it something we do casually; we drench ourselves in smells, we wallow in them. Not only do we perfume our bodies and homes, we perfume almost every object that enters our lives, from our cars to our toilet paper. Used-car dealers have a "new-car" spray, guaranteed to make a buyer feel good about the oldest tin warthog. Real estate dealers sometimes spray "cake-baking" aromas around the kitchen of a house before showing it to a client. Shopping malls add "pizza smell" to their air-conditioning system to put shoppers in the mood to visit their restaurants. Clothing, tires, magic markers, and toys all reek with scent. One can even buy perfume discs that play like records, except that they exude scent. As has been proven in many experiments, if you hand people two cans of identical furniture polish, one of which has a pleasant odor, they will swear that the pleasantly scented one works better. Odor greatly affects our evaluation of things, and our evaluation of people. Even so-called unscented products are, in fact, scented to mask the chemical odors of their ingredients, usually with a light musk. In fact, only 20 percent of the perfume industry's income comes from making perfumes to wear; the other 80 percent comes from perfuming the objects in our lives. Nationality influences fragrances, as many companies have discovered. Germans like pine, French prefer flowery scents, Japanese like more delicate odors, North Americans insist on bold smells, and South Americans want even stronger ones. In Venezuela, floor-cleaning products contain ten times as much pine fragrance as those in the United States. What almost all nationalities share is the need to coat our floors and walls with pleasant odors, especially with the smell of a pine forest or lemon orchard, to nest in smells.

A small shop on Third Avenue near Gramercy Park, like many such places throughout New York, sells a mélange of sensory delights. There are many pieces of Port Meiron china emblazoned with colorful, precisely detailed botanical drawings. Stationery and wrapping paper is all handmade, the woody fibers and imperfections thickly visible. Some are coarse-grained, with tutti-frutti splotches of color. The nose leads the way. Small bath-oil beads claim to smell like "Spring Rain" or "Nantucket." What does Spring Rain smell like? It's a popular scent. But would even the diehard sensuist know the difference between spring rain and, say, summer or fall rain?

Appealing first to the imagination, it puts a picture of spring rain in the mind, then you inhale its sweet mineral essence and think, perhaps, of the red-capped lichens called "British soldiers" you discovered in the Berkshires when you were ten. Or remember the scent of rain on the olive-drab tent, and hear the rain falling on canvas like a thousand drumming fingers. Gramercy Park seems only a small eddy in time from those distant years. One shelf in the store is devoted entirely to environmental fragrances. "Use with our aluminum light bulb ring to perfume your living spaces" one of the packages explains. *Parfum de l'Ambiance.* Tint the air with scent, perfume what enters your nostrils, bathe in sweetness while you walk from one room to another, stir the fragrance by dancing.

We seem unable to live in Nature without taking on its smells and wearing them as talismans, imagining we possess their ferocity, magnetism, or zest. On the one hand, we live in quarters sanitary and orderly, and if Nature should be rude enough to enter—in the form of a vole, fly, or termite crawling along the skirting boards, or a squirrel in the foundations, or a bat in the attic—we stalk it with the blood lust of a hunter. On the other hand, we insist on bringing Nature indoors with us. We touch the wall and make daylight flood a room, we turn a dial and it's summer, we surround ourselves with a caravan of completely unnecessary outdoor smells—pine, lemon, flowers. We may not need smell to survive, but without it we feel lost and disconnected.

ANOSMIA

One rainy night in 1976, a thirty-three-year-old mathematician went out for an after-dinner stroll. Everyone considered him not just a gourmet but a wunderkind, because he had the ability to taste a dish and tell you all its ingredients with shocking precision. One writer described it as a kind of "perfect pitch." As he stepped into the street, a slow-moving van ran into him and he hit his head on the pavement when he fell. The day after he got out of the hospital, he discovered to his horror that his sense of smell was gone.

Because his taste buds still worked, he could detect foods that were salty, bitter, sour, and sweet, but he had lost all of the heady succulence of life. Seven years later, still unable to smell and deeply depressed, he sued the driver of the van and won. It was understood, first, that his life had become irreparably impoverished and, second, that without a sense of smell his life was endangered. In those seven years, he had failed to detect the smell of smoke when his apartment building was on fire; he had been poisoned by food whose putrefaction he couldn't smell; he could not smell gas leaks. Worst of all, perhaps, he had lost the ability of scents and odors to provide him with heart-stopping memories and associations. "I feel empty, in a sort of limbo," he told a reporter. There was not even a commonly known name for his nightmare. Those without hearing are labeled "deaf," those without sight "blind," but what is the word for someone without smell? What could be more distressing than to be sorely afflicted by an absence without a name? "Anosmia" is what scientists call it, a simple Latin/Greek combination: "without" + "smell." But no casual term—like "smumb," for instance—exists to give one a sense of community or near-normalcy.

The "My Turn" column in *Newsweek* of March 21, 1988, by Judith R. Birnberg, contains a deeply moving lament about her sudden loss of smell. All she can distinguish is the texture and temperature of food. "I am handicapped: one of 2 million Americans who suffer from anosmia, an inability to smell or taste (the two senses are physiologically related). . . . We so take for granted the rich aroma of coffee and the sweet flavor of oranges that when we lose these senses, it is almost as if we have forgotten how to breathe." Just before Ms. Birnberg's sense of smell disappeared, she had spent a year sneezing. The cause? Some unknown allergy. "The anosmia began without warning. . . . During the past three years there have been brief periods—minutes, even hours—when I suddenly became aware of odors and knew that this meant that I could also taste. What to eat first? A bite of banana once made me cry. On a few occasions a remission came at dinner time, and my husband and I would dash to our favorite restaurant. On two or three occasions I savored every miraculous mouthful through an entire meal. But most times my taste would be gone by the time we parked the car." Although there are centers for treating smell and taste dysfunction (of which Monell is probably the best known), little can be done about anosmia. "I have had a CAT scan, blood tests, sinus cultures, allergy tests, allergy shots, long-term zinc therapy, weekly sinus irrigations, a biopsy, cortisone injections into my nose and four different types of sinus surgery. My case has been presented to hospital medical committees. . . . I have been through the medical mill. The consensus: anosmia caused by allergy and infection. There can be other causes. Some people are born this way. Or the olfactory nerve is severed as a result of concussion. Anosmia can also be the result of aging, a brain tumor or exposure to toxic chemicals. Whatever the cause, we are all at risk in detecting fires, gas leaks and spoiled food." Finally, she took a risky step and allowed a doctor to give her prednisone, an anti-inflammatory steroid, in an effort to shrink the swelling near olfactory nerves. "By the second day, I had a brief sense of smell when I inhaled deeply. . . . The fourth day I ate a salad at lunch, and I suddenly realized that I could taste everything. It was like the moment in 'The Wizard of Oz' when the world is transformed from black and white to Technicolor. I savored the salad: one garbanzo bean, a shred of cabbage, a sunflower seed. On the fifth day I sobbed—less from the experience of smelling and tasting than from believing the craziness was over."

At breakfast the next day, she caught her husband's scent and "fell on him in tears of joy and started sniffing him, unable to stop. His was a comfortable familiar essence that had been lost for so long and was now rediscovered. I had always thought I would sacrifice smell to taste if I had to choose between the two, but I suddenly realized how much I had missed. We take it for granted and are unaware that *everything* smells: people, the air, my house, my skin. . . . Now I inhaled all odors, good and bad, as if drunk." Sadly, her pleasures lasted only a few months. When she began reducing the dosage of prednisone, as she had to for safety's sake (prednisone causes bloating and can suppress the immune system, among other unpleasant side effects), her ability to smell waned once more. Two new operations followed. She's decided to go back on prednisone, and yearns for some magical day when her smell returns as mysteriously as it vanished.

Not everyone without a sense of smell suffers so acutely. Nor are all smell dysfunctions a matter of loss; the handicap can take strange forms. At Monell, scientists have treated numerous people who suffer from "persistent odors," who keep smelling a foul smell wherever they go. Some walk around with a constant bitter taste in their mouths. Some have a deformed or distorted sense of smell. Hand them a rose, and they smell garbage. Hand them a steak and they smell sulfur. Our sense of smell weakens as we get older, and it's at its peak in middle age. Alzheimer's patients often lose their sense of smell along with their memory (the two are tightly coupled); one day Scratch-and-Sniff tests may help in diagnosis of the disease.

Research done by Robert Henkin, from the Center for Sensory Disorders at Georgetown University, suggests that about a quarter of the people with smell disorders find that their sex drive disappears. What part does smell play in lovemaking? For women, especially, a large part. I am certain that, blindfolded, I could recognize by smell any man I've ever known intimately. I once started to date a man who was smart, sophisticated, and attractive, but when I kissed him I was put off by a faint, cornlike smell that came from his cheek. Not cologne or soap: It was just his subtle, natural scent, and I was shocked to discover that it disturbed me viscerally. Although men seldom report such detailed responses to their partner's natural smell, women so often do that it's become a romantic cliché: When her lover is away, or her husband dies, an anguished woman goes to his closet and takes out a bathrobe or shirt, presses it to her face, and is overwhelmed by tenderness for him. Few men report similar habits, but it's not surprising that women should be more keenly attuned to smells. Females score higher than males in sensitivity to odors, regardless of age group. For a time scientists thought estrogen might be involved, since there was anecdotal evidence that pregnant women had a keener sense of smell, but as it turned out prepubescent girls were better sniffers than boys their age, and pregnant women were no more adept at smelling than other women. Women in general just have a stronger sense of smell. Perhaps it's a vestigial bonus from the dawn of our evolution, when we needed it in court-ship, mating, or mothering; or it may be that women have tradition-ally spent more time around foods and children, ever on the sniff for anything out of order. Because females have often been responsible for initiating mating, smell has been their weapon, lure, and clue.

PRODIGIES OF SMELL

Just as there are people with distorted, failing, or nonexistent senses of smell, there are those at the other end of the olfactory spectrum, prodigies of the nose, the most famous of whom is probably Helen Keller. "The sense of smell," she wrote, "has told me of a coming storm hours before there was any sign of it visible. I notice first a throb of expectancy, a slight quiver, a concentration in my nostrils. As the storm draws near my nostrils dilate, the better to receive the flood of earth odors which seem to multiply and extend, until I feel the splash of rain against my cheek. As the tempest departs, receding farther and farther, the odors fade, become fainter and fainter, and die away beyond the bar of space." Other individuals have been able to smell changes in the weather, too, and, of course, animals are great meteorologists (cows, for example, lie down before a storm). Moist-ening, misting, and heaving, the earth breathes like a great dark beast. When barometric pressure is high, the earth holds its breath and vapors lodge in the loose packing and random crannies of the soil, only to float out again when the pressure is low and the earth exhales. The keen-nosed, like Helen Keller, smell the vapors rising from the soil, and know by that signal that there will be rain or snow. This may also be, in part, how farm animals anticipate earth-quakes—by smelling ions escaping from the earth.

People dressing for a dinner party on a stormy night won't need to use as much perfume, because perfume smells strongest just before a storm, in part because moisture heightens our sense of smell, and in part because the low pressure makes a fluid as volatile as perfume spread even faster. After all, perfume is 98 percent water and alcohol, and only 2 percent fat and perfume molecules. At times of low pressure molecules evaporate faster, and can waft from one's body into the alcoves of a room at considerable speed. This is also true, even on sunny days, in high-elevation cities such as Mexico City, Denver, or Geneva, where barometric pressures are always low because of the altitude. The ideal time and place to overwhelm a restaurant with one's new perfume would be at the 7,000-feet-high El Tovar Lodge, perched right on the sense-staggering edge of the Grand Canyon, when a storm is brewing.

Helen Keller had a miraculous gift for deciphering the fragrant palimpsest of life, all the "layers" that most of us read as a blur. She recognized "an old-fashioned country house because it has several layers of odors, left by a succession of families, of plants, of perfumes and draperies." How someone blind and deaf from birth could understand so well the texture and appearance of life, let alone the way our eccentricities express themselves in the objects we enjoy, is one of the great mysteries. She found that babies didn't yet have a "personality scent," unique odors she could identify in adults. And her sensuality expressed itself in smell—and explained an age-old attraction: "Masculine exhalations are, as a rule, stronger, more vivid, more widely differentiated than those of women. In the odor of young men there is something elemental, as of fire, storm, and salt sea. It pulsates with buoyancy and desire. It suggests all the things strong and beautiful and joyous and gives me a sense of physical happiness."

A FAMOUS NOSE

Those people with the nimblest sense of smell often end up working for perfumeries; some, if they are also imaginative and daring, create the great perfumes. In a sea of flowers, roots, animal secretions, grasses, oils, and artificial smells, they must be able to remember thousands of ingredients available to a perfumer, and the alchemical ways to blend them. They need an architect's sense of balance and a bookie's cunning. These days, laboratories can mimic natural essences, which is just as well, since we don't have reliable natural extracts of such flowers as lilac, lily of the valley, or violet. But to produce a persuasive rose oil may mean mixing five hundred ingredi-ents. On Fifty-seventh Street off Tenth Avenue in New York City, International Flavors and Fragrances Inc. houses the best profes-sional noses in the world. People in the business know the place simply as "IFF," a prolonged *if*, almost a wh*iff*, mecca for any company needing a smell. Although they create almost all of the expensive, lavishly advertised perfumes that appear in the depart-ment stores each season, and many of the flavors and smells we enjoy in everything from canned soup to kitty litter, they do their work anonymously. But they're the ones who provided the smell for a golf magazine's highly successful ad (peel away a paper golf ball and the smell of freshly mown grass surges up to your nostrils), as well as an amusement park's "cave" odor, and the habitat smells of New En-gland woodlands, African grasslands, Samoa, and other locales for displays in the American Museum of Natural History. Turning a fake Christmas tree into a Tyrolean pine forest in the mind of the inhaler is no problem. In fact, that's one of their simplest tricks. They are sensuous ghostwriters, inventors of rapture, creating the gold-plated aromas that influence and persuade us, without our knowing it. Eighty percent of men's colognes are created in their laboratories, and nearly that much of women's. Though they refuse to name names, in their hallways glass cases display perfumes by Guerlain, Chanel, Dior, Saint Laurent, Halston, Lagerfeld, Estée Lauder, and many others, to which they gave birth. Some of their noses point at computer consoles, others are at work in rooms clut-tered with papers and bottles. To them falls the ultimate paradox of creating a perfume that, on the one hand, is innovative, fresh, and exciting, and, on the other, is not too brazen or bizarre, but accept-able to large numbers of people. Scent strips, or Scratch-and-Sniff strips, have made their work easier to share. Pick up a magazine these days, and you'll be assaulted by pages that smell of a Rolls-Royce's leather upholstery, or of lasagne, or even of a new perfume. Invented at 3M Corporation only a decade ago, the strips contain microscopic balls full of fragrance. When you scratch, or tear back the flap, the balls rip open and the scent rushes out. Giorgio was the first com-pany to advertise their perfume with scent strips. Now it's difficult to find a magazine that doesn't smell. I have on my desk right now a collection of over forty scent strips advertising perfumes, with slogans—for Estée Lauder's Knowing, "Knowing is all"; Liz Claiborne's feminist "All you have to be is you" for her signature fragrance; Parfums Fendi's "La passione di Roma," in which a marble-cheeked young girl is caught passionately kissing a statue; Yves Saint Laurent's Opium is minus any verbal slogan, but its accompanying photograph of a beautiful woman in a gold-lamé suit, lying half dead in an opium delirium on a bed of orchids, makes its own perverse statement. There are thirty odor evaluators at IFF, on call to smell about a hundred fragrances a day. One spring afternoon, I meet their brilliant nose Sophia Grojsman, a robustly alive, Russian-born woman. Her short black hair is held back with a navy-and-white-striped headband. Her blue eyeshadow vibrates over dark lively eyes; she wears bright red nail polish and a denim suit with silver zippers. For a world-class nose on a deadline she seems relaxed and alert at the same time, as she sprawls behind her cluttered desk, right in the middle of which is a small trio of the monkeys who represent see-no-evil, speak-no-evil, hear-no-evil. Smell-no-evil doesn't rate a monkey.

"When did you first know that you had a special nose?"

"When I was a child in Russia, there were gigantic fields of flowers all around the little town where I lived." She smiles as she says it, and her eyes drift for a moment; the memory obviously carries her back forty years. "And there was an enormous amount of odor everywhere. The sky was thick with smells. I was always picking flowers . . ."

An abrupt knock at the door. A young woman walks in briskly, her long thin bare arms extended. "If you could smell me?" she says to Sophia. Sophia gets up and takes the woman's left arm first—the warmer arm, because it's nearer the heart—and presses her nose close and sniffs at the wrist and then again at the elbow. Then she sniffs twice at the other arm.

"What do you think?" Sophia asks me.

I sniff the arms. "Lovely."

"But in which order?"

The scents are so light, so quiet against my nose that it's hard to think of them as four distinct smells with individual personalities to be ranked. In one scene in *Bus Stop,* Marilyn Monroe sits in a diner, playing with two peas on her plate, choosing a favorite. There is always something about one that's better than another, she tells her companion; you can always choose. For me, life offers so many complexly appealing moments that two beautiful objects may be equally beautiful for different reasons and at different times. How can one choose? Still, here, on the extended arms, there is no doubt about number one—a slightly musky, basically floral scent at the woman's left wrist. Second? A lighter version of it at her left elbow. The smell on the right arm seems a shade fruitier, though somewhat attractive. I tell Sophia, who nods her head knowingly.

"Those are the two versions we need to work on," she says. A lab technician appears at a sliding glass window between her and the shelves upon shelves of bottles holding natural and synthetic essences, a real magician's larder. "I need the H formula," Sophia says to the technician, who returns to her cupboards. Sophia leans back in her chair and makes a gesture with her hands as if throwing confetti into the air. "This is a total madhouse today. We've had an emergency that I'm trying to attend to."

A scent emergency? What on earth could that be? When I ask, Sophia remains sphinxlike. In this corporate world, formulae and everything related to them are guarded and double-guarded. The people who blend the final fragrances don't know what they're blending; the ingredients and the batches carry only code numbers.

"We lived right at the end of the little town," Sophia says, returning to her memories, "and there were lilac bushes and whole fields of narcissus and violets. A world of natural smells was all around, a part of Russia that wasn't badly destroyed. As a child, I would wander off into the fields; I was desperately curious, snooping into everything. This was postwar time, and there weren't many children. I was surrounded by grown-ups and I would wander off by myself and pick up and smell the moss, the twigs, the leaves."

"When you're creating a scent, what is the process?" I ask, remembering that one of the great perfumers said he got his ideas from dreams, another that he kept a diary of everything he smelled when he traveled.

"You always have an image in your head. You can actually smell the accords, which are like musical chords. Perfumery is closely related to music. You will have simple fragrances, simple accords made from two or three items, and it will be like a two- or three-piece band. And then you have a multiple accord put together, and it becomes a big modern orchestra. In a strange way, creating a fragrance is similar to composing music, because there is also a similarity in finding the 'proper' accords. You don't want anything being overpowering. You want it to be harmonious. One of the most important parts of putting a creation together is harmony. You could have layers of notes coming through the fragrance, but yet you still feel it's pleasing. If the fragrance is not layered properly, you'll have parts and pieces sticking out, it will make you uncomfortable, something will disturb you about it. A fragrance that's not well balanced is not well accepted."

"Do you have the smells grouped in your mind and memory, the way woodwinds occupy one element of an orchestra and strings another?"

"Yes, but most of what I've created has come from totally abstract floral accords that just came along—once I'd got them, I looked for other parts and pieces to go along with them. First there is the inspiration, then ways to revise it until I've finally got what I'm after. I prefer very flowery, very feminine accords. I'm better at female fragrances than male fragrances, although I've done both. I've also made functional products—"

"Like the scents for soaps, cleaners, polishes, paper products, and so on?"

"Exactly. But those things are easy and quick to do. If I'm trying to create the next best perfume in the world, well . . . it takes longer."

"One of the company's officials told me that you've 'made some of the most famous perfumes in the world known to man or beast,' but that you're not going to tell me which they are."

"We can't tell." She pulls a long brown cigarette from a pack that says MORE and lights it.

"Does smoking affect your nose?"

"I'm sure it does something, but this is my environment, so I'm used to it. It's just one of the usual smells in my world."

"Do you protect your nose; are you hyper-concerned about it?"

"Not at all. I'm really very casual. Naturally, I don't want to get sick: It's frustrating to have a blocked-up nose, very hard for a perfumer to work in that condition."

"When you walk around the city, are you more acutely aware of smells than other people?"

"You know, it's a funny thing—an incredible phenomenon—but, because I work a lot, sometimes long hours, when I walk out of the building a little switch in my brain turns me off and I don't smell anything at all. In fact, there could be something burning on the stove at home and I wouldn't smell it! My husband says: 'You're a perfumer and you can't smell the burning!' My brain turns off totally.

"But I find myself tuning in to people at odd moments. Sometimes someone kisses you and you recognize their individual smell. There's a certain smell to a baby's skin, to the top of a baby's head. Men do this less than women. Some people naturally smell 'sexy.' If I had to describe it," she says, wafting the cigarette like a censer, as she searches for the precise description, "I'd call it a very delicate, ambery-musky accord. I use a lot of it in my fragrances.

"There are certain accords that every perfumer uses. But you can recognize someone's handwriting, so to speak, by smelling a fragrance. Other perfumers can recognize my work, as I can theirs. They smell a new perfume and they say: Ah, this is Sophia's, that's Jenny's, and so on. They know the signatures."

"I was in Saks last week," I explain, "on a smell safari, and I noticed that the trend seems to be for perfumes with names that suggest danger, prohibited substances, neuroses, and so on. . . ." I said that merchandisers seem to prefer smells that conjure up comfort and security, love and romance, but name them Decadence, Poison, My Sin, Opium, Indiscretion, Obsession, Tabu. In addition to the popular designer names and the bottled mystique of the superstars, they offer illegal substances and warnings. A woman may dress demurely, but in her mind and on her pulse-points she is as addictive as Opium, as dangerous as Poison, the cause for Obsession, expert in the ways of love so enthralling they're Tabu, ready for hedonistic Decadence, worth any Indiscretion, even transgressing the laws of God, as in Sin.

"Yes, but if you look at them closely, you discover that they're all based on certain classic scents, they're simply new interpretations of those classics. There are many instant successes, but true classics last over a decade. Chanel No. 5 was created in the early 1920s and still sells very well. Opium is nothing new. The mother of Opium is Youth Dew, which is about thirty years old. It's a variation on it, that's all, and it's also related to Cinnabar. If you smell the three together, you'll see."

"So, using your metaphor of music, a new fragrance is often a variation on an established theme?" She nods.

"Do *you* wear perfumes?"

"Not when I come to work. I do wear a lot of experiments. As I work with it, I wear it. I like to get the reaction of people to what I'm wearing. They're good judges. I was working on one fragrance, and when I walked out onto Fifty-seventh Street, I was followed by a drunk and I got scared. I started to run away from him, and he said: 'Lady, don't run. The perfume is so beautiful, I was following the perfume.' It turned out to be a winner."

"Since the beginning of time, people have perfumed themselves. Doesn't that seem an odd thing to do? To put flowers, fruits, and animal secretions on your body? Why do we do it?"

"Ah," she says, tossing her fingers as if setting free a handful of butterflies, "when I first saw Picasso's *Guernica,* it was disturbing. I was horrified and fascinated at the same time. It was disturbing, but also deeply moving. Perfumes do that, too—shock and fascinate us. They disturb us. Our lives are quiet. We like to be disturbed by delight.

"One of the most gratifying experiences for me," she says unexpectedly, "was when I made a functional product, the smell for a detergent. I was walking along the street, and there were two old ladies buying a newspaper. I said, 'Oh, ladies, you washed your clothes in so-and-so.' They said, 'How on earth did you know?' I said, 'I can smell it.' They were so happy and so was I, because these ladies can't afford a two- or three-hundred-dollar perfume but they can afford a detergent, and they were happy that it smelled good. I was pleased that I touched a portion of humanity that could never be able to afford the perfumes you just smelled here."

"How lucky you are to be able to spend your life in this way, creating scents that will make women feel good about themselves."

"Sometimes there are grueling hours of work. A perfumer's life is not a picnic. It's not what it used to be. In the great old days, there were perfumers who were free-lancers. A famous perfumer would make one fragrance in three or four years, and they had no restrictions—no price limit, no deadline. They would make two or three experiments a day for perhaps a week, then really *live* with it, wear it for weeks and weeks without any pressure. What's happening now is that it's very commercialized. You want to do things that will make a name for you, money for the company, and you must do them fast. A perfume can't be made overnight. Every perfumer has little accords that, during their ten years of practice, they put away and keep in their memory bank. Oh, I need a floral, they might say, I remember that floral I had years ago. But it must be new. You'd be a fool to sell a copy. You can't plagiarize. You have to start from scratch. But there are accords you might return to as themes, as a kind of shortcut. I make approximately five hundred to seven hundred formulas a year. Maybe you see two big pieces of business come out of that, but this doesn't mean all the seven hundred formulas aren't good."

"Doesn't it break your heart if you create a formula that really stirs you, but the customer doesn't care for it?"

She rolls her eyes and her face keens. "Of course, and it certainly does happen. I always try to make it work somewhere eventually, so that somebody finally gets it. You have to believe in the fragrance, believe that it will prevail, that it will be there sometime, somehow. I'm very persistent. I keep going back to it, rethinking it.

"There's something that I made recently and I can't tell you the name of it, but the fragrance is an *experience*. Wearing it is an experience. I happen to love it. The main accord of the fragrance started a while ago with one accord that I called "cleavage"—"headless," "bottomless," I have all these crazy names that I privately call things—and what cleavage smells to me like is a young woman's skin here"—she lifts her hands to show the area between the chin and the bosom—"There's something very sensual and sensational about this accord."

She takes a long paper tester and dips it into an amber bottle full of oil, hands it to me. As I waft the smell under my nose, sherbety flowers drift over my senses. It is a very young smell, girlish and innocent, full of soft ruffles and lightly talced skin.

"This is simple but very complicated-smelling. It says in a strange way 'Hug me.' It's a sexy note that men adore. I knew I had a winner when I made this." She hands me another dipstick, this one fresher and slightly more alive. "Now this is the perfume it became. The first oil was the skeleton. This is the result. From the first bottle, it went all the way down the line to the finished perfume. It's basically a floral, but the more you smell it the more delicate it becomes."

"Which is the most sensual perfume you've created?"

"This is an interesting question, because what's sexy and sensual for one isn't necessarily for the other. To me, this one is sensuous, not sexy, but sensuous."

"How about one that's vampy?"

"Try this one."

She hands me a new tester; I hold it under my nose and have a powerful response. I can taste something thick and amber, like butterscotch, on the back of my tongue. It has a thin vinyl covering to it and a fizzy muskiness seems to be coming up all around it in a halo. It smells deeply luscious. "What is it?" I ask, scrunching up my face in the automatic contortion of pleasure.

"It's basically a Shalimar-type formula. It's not on the market yet."

"Unlike the other one I sampled—'cleavage'—when I smell this I have a strong physical response. I can taste it."

She laughs. "Yes, that's what people say about my perfumes, that you can taste them. I'm very passionate about everything I do. I want my creations to stir your taste and smell and emotions all at once."

"Can you picture a perfume that you can't create? Is there an ideal form that you strive for?"

"Oh, I would like to make a perfume some day so seductive to men that no woman could be resisted. It would be the most incredible thing I could do in my life. This is not a professional feeling. It's strictly a female feeling."

"The whole world would become unsafe."

"Yes!" she says with relish.

"Let me know when you find it. I'll be your first guinea pig."

"*I'll* be my first guinea pig."

AN OFFERING TO THE GODS

When I leave IFF with its carnival of new smells and Fortune 500 status and its secret corridors that merge, veer off, and interflow like the workings of smell itself, I step outside into an atmosphere low-slung and broody. Steam rises from the manholes, as if there were one large sweat gland under the city. How does a professional nose stay acute in a city of warring smells, some of which are caustic? Perfumers aren't the only professional noses who must survive this urban sump. Doctors have always relied on their sense of smell, along with those of sight, feel, and hearing, to diagnosis diseases, especially in the days before sophisticated technology. Typhus is said to smell of mice; diabetes of sugar; the plague of mellow apples; measles of freshly plucked feathers; yellow fever of the butcher shop; nephritis of ammonia.*

We not only need all our senses, we need more of them, new senses. And, if necessary, we're willing to create and employ them outside our bodies, as scanning electron microscopes, radio telescopes, atomic scales. But we cannot do this effectively with smell. If smell is a relic, it's of a time of great intensity, need, instinct, and delirium, a time when we moved among the cycles of Nature as one

---

*Among the curious diseases recognizable by smell is maple syrup urine disease, which afflicts infants. Doctors aren't sure what produces the odor. The smell of acetone on a patient's breath often signals diabetes. "Menses breath" (some women develop an oniony smell) comes from a change in sulfur compounds in the body during a woman's menstrual cycle.

of its promising protégés. Except to taste and to scout danger, we don't really need smell any longer, but we will not let go. We will not be weaned. Evolution keeps trying to tug it gently from our hands, pull it away while we are sleeping, like a stuffed animal or favorite blanket. We cling to it tighter than ever. We don't want to be cut off from the realms of Nature that survive by smell. Most of what we do smell is accidental. Flowers have scents and bright colors as sex attractants; leaves have aromatic defenses against predators. Most of the spices, whose heady aromas we are drawn to, repel insects and animals. We are enjoying the plant's war machine. As one quickly learns in the Amazon rain forest, there is nothing wimpy about a plant. Because trees can't move to court each other or to defend themselves, they've become ingenious and aggressive about their survival. Some develop layers of strychnine or other toxic substances just under the bark; some are carnivorous; some devise flowers with intricate feather dusters to touch pollen to any bug, bird, or bat they have managed to lure with siren smells and colors. Some orchids mimic the reproductive parts of a female bee or beetle in order to trick the male into trying to copulate, so it will become dusted with pollen. One night a year, in the Bahamas, the *Selenicereus* cactus flowers ache into bloom, conduct their entire sex lives, and vanish by morning. For several days beforehand, the cactuses develop large pregnant pods. Then one night, awakened by a powerful smell of vanilla, you know what has happened. The entire moonlit yard is erupting in huge, foot-wide flowers. Hundreds of sphinx moths rush from one flower to another. The air is full of the baying of dogs, the loud fluttering of the moths that sounds like someone riffling through a large book, and the sense-drenching vanilla nectar of the flowers, which disappear at dawn, leaving the cactuses sated for another year.

In ancient times, when perfumes were almost as mystical as they were precious, explorers set out in search of their healing or aphrodisiac qualities. Our sense of smell has contributed to the spread of language, which evolved at the crossroads of ancient trade routes. Yearning for spices, perfumes, medicinal herbs, and exotic talismans, people set sail across continents and seas, and when they arrived they had to be able to haggle and, eventually, keep records. I don't recall anyone celebrating the senses of smell or taste during our bicentennial in 1976. But Columbus's quest, we tend to forget, was sensuous as well as capitalistic, adventuresome, and ego-driven. It was partly the obsessive demand for exotic spices and perfumes that prompted him to set sail in the first place.

Perfume began in Mesopotamia as incense offered to the gods to sweeten the smell of animal flesh burned as offerings, and it was used in exorcisms, to heal the sick, and after sexual intercourse. The word's Latin etymology tells us how it worked: *per* = through + *fumar* to smoke. Tossed onto a fire, incense would fill the sky with a smoke otherworldly and magical, which stung the nostrils as if clamorous spirits were clawing their way into the body. Perfumed smoke began with the things of this earth but climbed quickly into the realm of the gods. Atop the famous ziggurat-shaped Tower of Babel, which stretched closer to the gods than mortals could reach, priests lit pyres of incense. Given the general hand-me-down history of fashion and luxury, perfumes were probably reserved for the gods at first, then priests were allowed them, then godlike leaders, then leaders, then aides, all the way down the social totem pole. Prehistoric people applied perfumes to their bodies, as primitive (and more sophisticated) peoples do today. An anthropologist friend who works with Indian tribes in the Amazon tells of one tribe in which the women wrap a kind of skirt made of sage around their waists and the men rub a fragrant root under their arms as deodorant. The first civilization to go on record as using perfume regularly, extravagantly, and with nuance was Egypt. Their elaborate burial and embalming practices required spices and unguents. They burned tons of incense in elaborate worship rituals. Scent became a national obsession during the reign of Queen Hatshepsut, of the New Kingdom (1558–1085 B.C.), who planted large botanical gardens and burned incense on the terraces leading to her temples. The Egyptians used lavish quantities of perfume and incense in their religious cults, eventually coming to enjoy them for personal daily use as well, especially during Egypt's Golden Age. They anointed their bodies with perfumes to ward off magical hexes, for medicinal purposes, and as beauty lotions, because they prized the feel of silky, scented skin. Egyptians discovered enfleurage (pressing aromatics into fatty oils) and created beautiful glass vessels to hold their potions, including the *millefiori* and other styles Venetian glassmakers were to use centuries later; they indulged in elaborate beauty rituals and had an almost modern fascination with makeup. If we were to observe a woman of ancient Egypt fixing her face and hair before a dinner party, we would find her seated at her makeup table, which would hold a variety of elegant, imaginatively designed perfume spoons, receptacles for unguents, vases, flacons, and boxes of eye shadow. She might well have a tattoo of a scarab or flower on her shoulder—Egyptian women were fond of tattoos. (When an Egyptian tomb was opened in the 1920s and a mummy discovered to be delicately tattooed, Lady Randolph Churchill and other socialites decided to get scarab tattoos themselves.) An ancient Egyptian socialite attending a party would wear a wax cone of unguent on the top of her head; it would melt slowly, covering her face and shoulders with a trickle of perfumed syrup. It probably felt as if small beetles were crawling all over her, pushing balls of fragrance. The Egyptians were a clean, ingeniously sybaritic people obsessed with hygiene; they invented the sumptuous art of the bath—an art that might be restorative, sensuous, religious, or calming, depending on one's mood. This they would usually follow with a massage of aromatic oils to soothe the muscles and calm the nerves—aromatherapy, a technique first used in the embalming of mummies. Researchers at Yale's Psychophysiology Center are studying how smell can decrease stress and increase alertness. They claim that the smell of spiced apples can reduce blood pressure in people under stress and avert a panic attack, and that lavender can wake up one's metabolism and make one more alert. *The Chronicle of Higher Education* reports that related tests at the University of Cincinnati have shown how fragrances added to the atmosphere of a room can increase typing speed and work efficiency in general.

At the Sonesta Beach Spa in Bermuda, I stretch out on a table in front of a window, through which I can see and hear the crash and caterwaul of the sea. A pretty young woman with large blue eyes enters the small room, wearing a white belted cosmetician's dress. Fresh from Yorkshire, she hasn't been on the island long enough to develop a tan on the twelve weekends she's had free. Her boyfriend is in the marine division of the Bermuda police, and yesterday she went to the Cricket Cup Match with him. She has bunions on her feet, inherited from her father's side of the family, along with the small symmetrical nose she thinks is too large, and the straight blond hair she thinks is too thin. Today she has me lie on my back and then discreetly covers me with blue terry-cloth towels, which she will rearrange as the hour progresses. In the past few days, she has seen my body enough to know its flaws and graces. Only a lover could touch it more often, or better. Now we are as relaxed about my nakedness as old spouses. She explains the next treatment: aroma-therapy. This ancient Egyptian technique fell out of favor for many hundreds of years, reemerging in the eighteenth century, when aromatics and herbals returned to fashion. Because what I seek is relaxation more than mummification, my masseuse will blend laven-der, neroli, and sandalwood in a sweet almond-oil base and massage my body from head to toe in windblown patterns that concentrate on the lymph system. I am not to shower afterwards, because the oils massaged into circulation need time to penetrate and soothe. Starting at the calves, she massages in fan shapes, rolling, circular, roaming, always returning to the point of origin, then veering off again in symmetrical arcs or ripples. The fragrance—musky, heavy, Mideastern—seems to roll up my body. After the legs, she does the rump; then the back, pausing to apply pressure at certain stations down each side of the spine. She skates across the shoulder blades, probing, then smoothing. The treatment's effect comes in part, she quietly explains, from the "energy flow" created between the two bodies. A veil of scent rises around my neck, collars me in pungent mist; her hands keep revolving, heating the oils. Unexpectedly, my mind begins to drift to when I was a child and my father drove us to Florida all the way from Illinois for a brief summer vacation. The journey from outside Chicago to Florida was long, and my mother packed a cold chest of sandwiches and Hawaiian Fruit Punch, a wicker basket of our favorite toys and some new comic and activity books. I picture the trip in such surprising detail: the "yup-yup leaves" that fairies in one of the comics harvested, the Spanish moss on the trees we passed, my mother, who loved to sing in the car, sitting in a gray dress patterned with large, mauve, cabbagey roses. She wore her straight brown hair Ava Gardner style. Sometimes, when she was silent, her left index finger would move sharply in a way that intrigued me. I was too young to understand that she was probably talking to herself. Why have I remembered that time? I was eight. My mother had me when she was thirty. I am now the age she was then, and she had two children. This vivid memory stays with me and fills me with a thick, warm lager. Then the masseuse swaddles me in a pale-blue blanket. The light-blue walls of the room have a small woodblock print: thousands of brown chevrons. Above each one floats a pair of gray quotation marks angled like those at the end of an utterance.

CLEOPATRA'S HEIRS

Masters of aromatics, the Egyptians had many uses for cedarwood: in mummification, as incense, and to protect papyruses from the assaults of insects. Cleopatra's cedarwood ship, on which she received Antony, had perfumed sails; incense burners ringed her throne, and she herself was scented from head to toe. I return to her now because she was the quintessential devotee of perfume. She anointed her hands with *kyphi*, which contained oil of roses, crocus, and violets; she scented her feet with *aegyptium*, a lotion of almond oil, honey, cinnamon, orange blossoms, and henna. The walls were an aviary of roses secured by nets, and her regally perfumed presence arrived before her, like a kind of calling card in the scent-drenched wind. As Shakespeare imagines the scene: "From the barge/ A strange invisible perfume hits the sense/ Of the adjacent wharfs." Romans became famous for their spa-like grandeur, but they actually borrowed the bath from the sybaritic Egyptians.

In the ancient world, royal architecture itself was often aromatic. Potentates built whole palaces of cedarwood, in part because of its sweet, resiny scent, and in part because it was a natural insect repellent. In the Nanmu Hall at the imperial summer palace of the Manchu emperors at Ch'eng-te, the beams and paneling, all of cedarwood, were lacquerless and paintless, so that the fragrance of the wood could influence the air. Builders of mosques used to mix rose water and musk into mortar; the noon sun would heat it and bring out the perfumes. The doors of Sargon II's eighth-century B.C. palace in what is now Khorsabad were so scented that they would waft perfume when visitors entered or left. Pharaonic barges and coffins were made of cedarwood. The temple of Diana at Ephesus, one of the Seven Wonders of the ancient world, which had columns almost sixty feet high, survived for two hundred years, then burned down in 356 B.C., aromatically aflame. Legend says that, in shame or as an offering, it burned when Alexander the Great was born.

Ancient he-men were heavily perfumed. In a way, strong scents widened their presence, extended their territory. In the pre-Greek culture of Crete, athletes anointed themselves with specific aromatic oils before the games. Greek writers of around 400 B.C. recommended mint for the arms, thyme for the knees, cinnamon, rose, or palm oil for the jaws and chest, almond oil for the hands and feet, and marjoram for the hair and eyebrows. Egyptian men, attending a dinner party, would receive garlands of flowers and their choice of perfumes at the door. Flower petals would be scattered underfoot, so they could make a fragrant stir when guests trod on them. Statues at these banquets often spurted scented water from their several orifices. Before retiring, a man would crush solid perfume until it was an oily powder and scatter it onto his bed so that he could absorb its scent while he slept. Homer describes the obligatory courtesy of offering visitors a bath and aromatic oils. Alexander the Great was a lavish user of both perfumes and incense, and was fond enough of saffron to have his tunics soaked in its essence. Babylonian and Syrian men wore heavy makeup and jewelry, as well as laboriously arranged coiffures of tiny ringlets set with perfumed lotions. In ancient Rome, the passion reached such heights that both men and women took baths in perfume, soaked their clothes in it, and perfumed their horses and household pets. The gladiators applied scented lotions all over—a different scent for each area of the body—before they fought. And, like other Roman men and women, they used pigeon dung to bleach their hair. In their equivalent of a locker room, before a gory contest with a lion, crocodile, or man, they might have been talking rough, but their hands were applying sweet scents. Roman women applied scents to different parts of their bodies, just as Roman men did, and I imagine they spent some time deciding whether sandalwood feet and jasmine breasts went well with a neroli neck and lavender thighs. With Christianity came a Spartan devotion to restraint, a fear of seeming self-indulgent, and so men stopped wearing scents for a while. (Even so, a religious symbolism attached to favorite flowers and their scents. For example, the carnation was in favor because its smell resembles that of cloves, and cloves themselves resemble the nails that were driven into Christ's cross.) As John Trueman puts it in *The Romantic Story of Scent:* "The men of the ancient world were clean and scented. European men of the Dark Ages were dirty and unscented. Those of medieval times, and of modern times up to about the end of the 17th century, were dirty and scented. . . . Nineteenth-century men were clean and unscented." But men seldom wandered far afield from desirable scents. The crusaders returned from their travails wearing rose water. Louis XIV kept a stable of servants just to perfume his rooms with rose water and marjoram, to wash his shirts and other apparel in a stew of cloves, nutmeg, aloe, jasmine, orange water, and musk; he insisted that a new perfume be invented for him every day. At "The Perfumed Court" of Louis XV, servants used to drench doves in different scents and release them at dinner parties, to weave a tapestry of aromas as they flew around the guests. The Puritans did away with scents, but soon enough men took them up again.

An eighteenth-century woman's dressing called for elaborate preparations and a discerning nose: She wore sweet-smelling hair powder and scented makeup; her perfumed clothes were kept in an aromatic clothespress; she lavishly perfumed her body, and then soaked cotton pomanders in cologne to tuck into her bodice. Potpourris sat on her tables, scenting the room from their Chinese porcelain containers ("porcelain" is a word with a fascinating history, which leads back, through cowry shells, to the genitals of a female pig, which is obviously what its silky texture reminded them of). At midday, she changed into a fresh array of aromas equally overwhelming. And then again at evening. Napoleon's passion for luxury included his favorite cologne water, made of neroli and other ingredients, 162 bottles of which he ordered from his perfumer, Chardin, in 1810. After he washed, he liked to pour cologne over his neck, chest, and shoulders. Even on his most arduous campaigns, in his elaborately decorated tent he took time to choose rose- or violet-scented lotions, gloves, and other finery. During the Napoleonic Wars, British sea captains sent on to the Empress Josephine roses destined for her garden at Malmaison (where she had 250 varieties); couriers with new varieties of roses had immunity passing between England and France. Elizabeth I adored gloves scented with ambergris; she not only wore perfumed cloaks, she required that her courtiers be heavily scented, too, so that they might surround her sweetly when they moved. A patron of the arts, Elizabeth was single-handedly responsible for the glory of the Elizabethan theater and the well-being of many writers, Shakespeare included, and she relished her position at the center of sensory and artistic life. She was particularly fond of Sir Walter Raleigh, and so, it may be assumed, of the strawberry cologne he liked to wear. Elizabeth kept her pets doused in scent, and she wore a pomander (an apple rolled in cinnamon and dressed in cloves) to ward off the plague.

This scent obsession started long before. The first gift to the Christ Child was incense and, in the eleventh century, Edward the Confessor presented Westminster Abbey with a sacred and surprisingly imperishable relic—some of the original frankincense carried by the Magi. In India, the art of *abhyanga,* a musky rubdown of female elephants to increase their sexual attraction to male elephants, still exists. In the ancient courts of Japan, clocks burned a different incense every fifteen minutes, and geishas were paid by the number of scent sticks consumed. Perfumes have obsessed every culture and religion, but the ultimate promise is probably in the Koran: Those religious enough to go to heaven will find there voluptuous companions called houris (from the Arabic *haurā'*, dark-eyed woman), who will attend to every whim and invent new cravings, which they will then quench. The ultimate font of delights, they are not merely perfumed—according to the Koran, they are made entirely of sandalwood. They are pure smell, pure pleasure. How fitting. In a sense, the houris return us to that time, before thought, before sight, when smell was all we had to guide us down the dimly lit corridors of evolution.

Touch

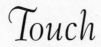

They are excessively warm hands,
that continually want to cool
themselves and involuntarily lay
themselves on any cold object,
outspread, with air between the fingers.
Into those hands the blood could
shoot, as it mounts to a person's
head, and when clenched,
they were indeed like the heads
of madmen, raging
with fancies.

Rainer Maria Rilke,
The Notebooks of Malte Laurids Brigge

THE FEELING BUBBLE

Our skin is a kind of space suit in which we maneuver through an atmosphere of harsh gases, cosmic rays, radiation from the sun, and obstacles of all sorts. Years ago, I read about a boy who had to live in a bubble (designed by NASA) because of the weakness of his immune system and his susceptibility to disease. We are all that boy. The bubble is our skin. But the skin is also alive, breathing and excreting, shielding us from harmful rays and microbial attack, metabolizing vitamin D, insulating us from heat and cold, repairing itself when necessary, regulating blood flow, acting as a frame for our sense of touch, aiding us in sexual attraction, defining our individuality, holding all the thick red jams and jellies inside us where they belong. Not only do we have unique fingerprints, we have unique pore patterns. According to Catholic belief, there is somewhere, protected in a secret vault, the relic foreskin of Christ. Since he ascended to heaven, his foreskin is the only mortal part of him that remains. We like to decorate our skin whenever we get the chance, and that is made easier by skin being portable, washable, and sloughy. Psychiatrist David Hellerstein's description of skin in *Science Digest* (September 1985) offers a simple, convenient picture of it in cross section:

> Skin is basically a two-layered membrane. The lower, thick spongey dermis, one to two millimeters thick, is primarily connective tissue, rich in the protein collagen; it protects and cushions the body and houses hair follicles, nerve endings and sweat glands, blood and lymph vessels. The upper layer, the epidermis, is 0.07 to 0.12 millimeter thick. It is primarily composed of squamous, or

scalelike, epithelial cells, which begin their lives round and plump at the boundary of the dermis and over a 15-to-30-day period are pushed upward, toward the surface, by new cells produced below. As they rise, they become flattened, platelike, lifeless ghosts, full of protein called keratin, and finally they reach the surface, where they are ingloriously sloughed off into oblivion.

Our skin is what stands between us and the world. If you think about it, no other part of us makes contact with something not us but the skin. It imprisons us, but it also gives us individual shape, protects us from invaders, cools us down or heats us up as need be, produces vitamin D, holds in our body fluids. Most amazing, perhaps, is that it can mend itself when necessary, and it is constantly renewing itself. Weighing from six to ten pounds, it's the largest organ of the body, and the key organ of sexual attraction. Skin can take a startling variety of shapes: claws, spines, hooves, feathers, scales, hair. It's waterproof, washable, and elastic. Although it may cascade or roam as we grow older, it lasts surprisingly well. For most cultures, it's the ideal canvas to decorate with paints, tattoos, and jewelry. But, most of all, it harbors the sense of touch.

The fingertips and tongue are much more sensitive than the back. Some parts of the body are ticklish, and others respond when we itch, shiver, or get gooseflesh. The hairiest parts of the body are generally the most sensitive to pressure, because there are many sense receptors at the base of each hair. In animals, from mice to lions, the whiskers around the mouth are extraordinarily sensitive; our body hairs are sensitive, too, but to a lesser degree. The skin is also thinnest where there's hair. Feeling doesn't take place in the topmost layer of skin, but in the second layer. The top layer of skin is dead, sloughs off easily, and contributes to that ring around the bathtub. This is why safecrackers are sometimes shown sandpapering their fingertips, making the top layer of skin thinner so that the touch receptors will be closer to the surface. A carpenter looking for rough patches may run a thumb over the plank of wood he has just planed. A cook may roll a bit of dough between a thumb and forefinger to test its consistency. Without having to look at the spot, we know at once where we cut ourself shaving, or where a stocking is starting to run. It's entirely possible to feel wet, even though we may not *be* wet (when washing dishes with plastic gloves on, say), which suggests the complex sensations that constitute touch. The reason it's easier to get our feet wet first when we brave an icy ocean is that there aren't as many cold receptors in the feet as there are on, for example, the tip of the nose.

In the Middle Ages, so-called witches and others who lived on the outskirts of the law, piety, or convention were burned at the stake. Mimicking the fire and brimstone of hell, it was the ultimate horror. Death would happen cell by cell, receptor by receptor; each of life's minute sensations would be torched. Today people who have some-how survived accidental burning come to the burn units of metropolitan hospitals to be re-dressed. If their burns are too deep for the body to repair by itself, they receive temporary coverings (cadaver skin, pigskin, lubricated gauze) until doctors can begin grafting skin from other body parts. Our skin makes up about 16 percent of our body weight (about six pounds), and stretches two square yards, but if too much of the body is burned, there isn't enough skin to graft.

In 1983, a Harvard Medical School team led by Dr. Howard Green found a revolutionary way to repair burned skin. Two small boys, Jamie and Glen Selby, were removing paint from their naked bodies when the solvent accidentally caught fire. Only five and six years old, the boys had burned themselves horrendously, one over 97 percent of his body, the other 98 percent. At the Shriners Burn Institute in Boston, doctors covered the boys with cadaver skin and artificial membrane, removed small squares of skin from their arm-pits and cultured them into large sheets of skin, which they grafted on gradually over a five-month period. They were able to repair half of the burned areas on each boy's body, and a little over a year later the boys went home to Casper, Wyoming. Although the boys didn't have any sweat glands or hair follicles on this skin, it was pliable and protective, and they were able to return to school. The doctors had been able to grow large quantities of new skin.

Here is how it is done: In a Harvard laboratory, doctors cut up a small patch of skin donated by a patient, treat it with enzymes, then spread it thinly onto a culture medium. After only ten days, colonies of skin cells begin linking up into sheets, which can then be chopped up and used to make further sheets. In twenty-four days, enough skin will be produced to cover an entire human body. The new skin is attached to gauze that has been saturated in Vaseline, then, gauze side up, sutured to the body. About ten days later, the gauze is removed, and the skin soon grows into a surface much smoother and more natural-looking than the rough one a normal skin graft usually leaves. As revolutionary as skin-growing is, other methods are equally intriguing. At New York Hospital—Cornell Medical Center, doctors have been experimenting with cadaver skin, which they grow in large quantities and store in a skin bank. At MIT, researchers have developed a high-speed technique that uses a quarter-sized patch of skin from the burn patient to manufacture a large amount of skin in under two hours. A graft can be made right away, without a three-week wait. In two weeks, the burn will be covered with fresh new skin. Again, the skin will lack hair follicles, sweat glands, and pigment, but it will protect and function like normal skin. Such techniques are not for minor burns or even small serious burns; they're useful only in patients who are severely burned over large areas and therefore have too little skin left for grafting. None of the techniques is without risk—delay; rejection; possible infection—but the very fact of being able to grow an organ, indeed the largest organ in the body, makes one pause to think about growing other organs or at least parts of them—eyes, ears, hearts—in a farm whose fields are pans and whose silos are test tubes.

SPEAKING OF TOUCH

Language is steeped in metaphors of touch. We call our emotions feelings, and we care most deeply when something "touches" us. Problems can be thorny, ticklish, sticky, or need to be handled with kid gloves. Touchy people, especially if they're coarse, really get on our nerves. *Noli me tangere,* legal Latin for "don't meddle or interfere," translates literally as "Don't touch me," and it was what Christ said to Mary Magdalen after the Resurrection. But it's also one term for the disease lupus, presumably because of the disfiguring skin ulcerations characteristic of that illness. A toccata in music is a composition for organ or other keyboard instrument in a free style. It was originally a piece intended to show touch technique, and the word comes from the feminine past participle of *toccare*, to touch. Music teachers often chide students for having "no sense of touch," by which they mean an indefinable delicacy of execution. In fencing, saying *touché* means that you have been touched by the foil and are conceding to your opponent, although, of course, we also say it when we think we have been foiled because someone's argumentative point is well made. A touchstone is a standard. Originally, touch-stones were hard black stones like jasper or basalt, used to test the quality of gold or silver by comparing the streaks they left on the stone with those of an alloy. "The touchstone of an art is its precision," Ezra Pound once said. D. H. Lawrence's use of the word touch isn't epidermal but a profound penetration into the core of someone's being. So much of twentieth-century popular dancing is simultaneous solo gyration that when people returned to dancing closely with partners again a couple of years ago, we had to call it something different—"touch dancing." "For a while there, it was touch and go," we say of a crisis or precarious situation, not realizing that the expression goes back to horse-and-carriage days, when the wheels of two coaches glanced off each other as they passed, but didn't snag; a modern version would be when two swerving cars brush fenders. What seems real we call "tangible," as if it were a fruit whose rind we could feel. When we die, loved ones swaddle us in heavily padded coffins, making us infants again, lying in our mother's arms before returning to the womb of the earth, ceremonially unborn. As Frederick Sachs writes in *The Sciences*, "The first sense to ignite, touch is often the last to burn out: long after our eyes betray us, our hands remain faithful to the world. . . . in describing such final departures, we often talk of losing touch."

FIRST TOUCHES

Although I am not a portly middle-aged gentleman with nothing else to do, I am massaging a tiny baby in a hospital in Miami. Often male retirees volunteer to enter preemie wards late at night, when other people have families to tend or a nine-to-five job to sleep toward. The babies don't care about the gender of those who cosset and cuddle them. They soak it up like the manna it is in their wilderness of uncertainty. This baby's arms feel limp, like vinyl. Still too weak to roll over by itself, it can flail and fuss so well the nurses have laid soft bolsters on its bed, to keep it from accidentally wriggling into a corner. Its torso looks as small as a deck of cards. That this is a baby boy lying on his tummy, who will one day play basketball in the summer Olympics, or raise children of his own, or become a heliarc welder, or book passage on a low-orbital plane to Japan for a business meeting, is barely believable. The small life form with a big head, on which veins stand out like river systems, looks so fragile, feels so temporary. Lying in his incubator, or "Isolette," as it's called, emphasizing the isolation of his life, he wears a plumage of wires— electrodes to chart his progress and sound an alarm if need be. Reaching carefully scrubbed, disinfected, warmed hands through the portholes of the incubator with pangs of protectiveness, I touch him; it is like reaching into a chrysalis. First I stroke his head and face very slowly, six times for ten seconds each time, then his neck and shoulders six times. I slide my hands down his back and massage it in long sweeping motions six times, and caress his arms and legs six times. The touching can't be light, or it will tickle him, nor rough, or it will agitate him, but firm and steady, as if one were smoothing a crease from heavy fabric. On a nearby monitor, two turquoise EKG and breath waves flutter across a radiant screen, one of them short and saw-toothed, the other leaping high and dropping low in its own improvisatory dance. His heartbeat reads 153, aerobic peak during a stiff workout for me, but calm for him, because babies have higher normal heart rates than adults. We turn him over on his back and, though asleep, he scrunches up his face in displeasure. In less than a minute, he runs a parade of expressions by us, all of them perfectly readable thanks to the semaphore of the eyebrows, the twisted code of the forehead, the eloquent India rubber of the mouth and chin: irritation, calm, puzzled, happy, mad. . . . Then his face goes slack and his eyelids twitch as he drifts into REM sleep, the blackboard of dreams. Some nurses refer to the tiny preemies, sleeping their sleep of the womb, as fetuses on the outside. What does a fetus dream? Gently, I move his limbs in a mini-exercise routine, stretching out an arm and bending the elbow tight, opening the legs and bending the knees to the chest. Peaceful but alert; he seems to be enjoying it. We turn him onto his tummy once more, and again I begin caressing his head and shoulders. This is the first of three daily touch sessions for him—it may seem a shame to interrupt his thick, druglike sleep, but just by stroking him I am performing a life-giving act.

Massaged babies gain weight as much as 50 percent faster than unmassaged babies. They're more active, alert, and responsive, more aware of their surroundings, better able to tolerate noise, and they orient themselves faster and are emotionally more in control. "Less likely to cry one minute, then fall asleep the next minute," as a psychologist, detailing the results of one experiment, explained in *Science News* in 1985, they're "better able to calm and console themselves." In a follow-up examination, eight months later, the massaged preemies were found to be bigger in general, with larger heads and fewer physical problems. Some doctors in California have even been putting preterms on small waterbeds that sway gently, and this experiment has produced infants who are less irritable, sleep better, and have fewer apneas. The touched infants, in these studies and in others, cried less, had better temperaments, and so were more appealing to their parents, which is important because the 7 percent of babies born prematurely figure disproportionately among those who are victims of child abuse. Children who are difficult to raise get abused more often. And people who aren't touched much as children don't touch much as adults, so the cycle continues.

A 1988 *New York Times* article on the critical role of touch in child development reported "psychological and physical stunting of infants deprived of physical contact, although otherwise fed and cared for . . . ," which was revealed by one researcher working with primates and others working with World War II orphans. "Premature infants who were massaged for 15 minutes three times a day gained weight 47 percent faster than others who were left alone in their incubators . . . the massaged infants also showed signs that the nervous system was maturing more rapidly: they became more active . . . and more responsive to such things as a face or a rattle . . . infants who were massaged were discharged from the hospital an average of six days earlier." Eight months later, the massaged infants did better in tests of mental and motor ability than the ones who were not.

At the University of Miami Medical School, Dr. Tiffany Field, a child psychologist, has been studying a group of babies admitted to the intensive care unit of its hospital for various reasons. With 13,000 to 15,000 births a year at the hospital, she never lacks for a steady supply of babies. Some are receiving caffeine for bradycardia and apnea problems, one is hydroencephalic, some are the children of diabetic mothers who must be carefully monitored. At one Iso-lette, a young mother sits on a black kitchen chair by her baby, reaches a hand in and gently strokes, whispering motherly nothings into its ear. Inside another Isolette, a baby girl wearing a white nightie with pink hearts bursts into a classic textbook wail that rises and pulses and sets off the alarm on her monitor. Across the room, a male doctor sits quietly beside a preemie, holding a two-pronged plastic stopper close to her nostrils, trying to teach her to breathe. Next to him, a nurse turns a baby girl onto her tummy and begins a "stim," as they call the massage, shorthand for stimulation. They use the word interchangeably as a verb or a noun. What old faces the preemies have! Changing expressions as they sleep, they seem to be rehearsing emotions. The nurse follows her massage schedule, stroking each part of the preemie six times for ten seconds. The stimulation hasn't changed the baby's sleep patterns, but she's been gaining thirty grams more a day and will soon be going home, almost a week ahead of what one would expect. "There's nothing extra going into the babies," Field explains, "yet they're more active, gain weight faster; and they become more efficient. It's amazing," she continues, "how much information is communicable in a touch. Every other sense has an organ you can focus on, but touch is everywhere."

Saul Schanberg, a neurologist who experiments with rats at Duke University, has found that licking and grooming by the mother rat actually produced chemical changes in the pup; when the pup was taken away from the mother, its growth hormones decreased. ODC (the "now" enzyme that signals it is time for certain chemical changes to begin) dropped in every cell in the body, and protein synthesis fell. Growth began again only when the pup was returned to the mother. When experimenters tried to reverse the bad effects without the mother, they discovered that gentle stroking wouldn't work, only very heavy stroking with a paintbrush that simulated the mother's tongue; after that the pup developed normally. Regardless of whether the deprived rats were returned to their mothers or stroked with paintbrushes by experimenters, they overreacted and required a great deal of touching, far more than they usually do, to respond normally.

Schanberg first began his rat experiments as a result of his work in pediatrics; he was especially interested in psychosocial dwarfism. Some children who live in emotionally destructive homes just stop growing. Schanberg found that even growth-hormone injections couldn't prompt the stunted bodies of such children to grow again, but tender loving care did. The affection they received from the nurses when they were admitted to a hospital was often enough to get them back on the right track. What's amazing is that the process is reversible at all. When Schanberg's experiments with infant rats produced identical results, he began to think about human preemies, who are typically isolated and spend much of their early life without human contact. Animals depend on being close to their mothers for basic survival. If the mother's touch is removed (for as little as forty-five minutes in rats), the infant lowers its need for food to keep itself alive until the mother returns. This works out well if the mother is away only briefly, but if she never comes back, then the slower metabolism results in stunted growth. Touch reassures an infant that it's safe; it seems to give the body a go-ahead to develop normally. In many experiments conducted all over the country, babies who were held more became more alert and developed better cognitive abilities years later. It's a little like the strategy one adopts on a sinking ship: First you get into a life raft and call for help. Baby animals call their mothers with a high-pitched cry. Then you take stock of your water and food, and try to conserve energy by cutting down on high-energy activities—growth, for instance.

At the University of Colorado School of Medicine, researchers conducted a separation experiment with monkeys, in which they removed the mother. The infant showed signs of helplessness, confusion, and depression, and only the return of its mother and continuous holding for a few days would help it return to normal. During separation, changes occurred in the heart rate, body temperature, brain-wave patterns, sleep patterns, and immune system function. Electronic monitoring of deprived infants showed that touch deprivation caused physical and psychological disturbances. But when the mother was put back, only the psychological disturbances seemed to disappear; true, the infant's behavior reverted to normal, but the physical distresses—susceptibility to disease, and so on—persisted. Among this experiment's implications is that damage is not reversible, and that the lack of maternal contact may lead to possible long-term damage.

Another separation study with monkeys took place at the University of Wisconsin, where researchers separated an infant from its mother by a glass screen. They could still see, hear, and smell each other, only touch was missing, but that created a void so serious that the baby cried steadily and paced frantically. In another group, the dividing screen had holes, so the mother and baby could touch through it, which was apparently sufficient because the infants didn't develop serious behavior problems. Those infants who suffered short-term deprivation became adolescents who clung to one another obsessively instead of developing into independent, confident individuals. When they suffered long-term deprivation, they avoided one another and became aggressive when they did come in contact, violent loners who didn't form good relationships.

In University of Illinois primate experiments, researchers found that a lack of touch produced brain damage. They posed three situations: (1) touch was not possible, but all other contact was, (2) for four hours out of twenty-four the glass divider was removed so the monkeys could interact, and (3) total isolation. Autopsies of the cerebellum showed that those monkeys who were totally isolated had brain damage; the same was true of the partially separated animals. The untampered-with natural colony remained undamaged. Shocking though it sounds, a relatively small amount of touch deprivation alone caused brain damage, which was often displayed in the monkeys as aberrant behavior.

As I rearrange the preemie in his glass home, I notice that on the walls a bright circus design shows clowns, a merry-go-round, tents, balloons, and a repeat banner that says "Wheel of Fortune." "Touch is far more essential than our other senses," I recall Saul Schanberg saying when we spoke, on Key Biscayne, at Johnson & Johnson's extraordinary conference on touch in spring, 1989, a three-day exchange of ideas that brought together neurophysiologists, pediatricians, anthropologists, sociologists, psychologists, and others interested in how touch and touch deprivation affect the mind and body. In many ways, touch is difficult to research. Every other sense has a key organ to study; for touch that organ is the skin, and it stretches over the whole body. Every sense has at least one key research center, except touch. Touch is a sensory system, the influence of which is hard to isolate or eliminate. Scientists can study people who are blind to learn more about vision, and people who are deaf or anosmic to learn more about hearing or smell, but this is virtually impossible to do with touch. They also can't experiment with people who are born without the sense, as they often do with the deaf or blind. Touch is a sense with unique functions and qualities, but it also frequently combines with other senses. Touch affects the whole organism, as well as its culture and the individuals it comes into contact with. "It's ten times stronger than verbal or emotional contact," Schanberg explained, "and it affects damn near everything we do. No other sense can arouse you like touch; we always knew that, but we never realized it had a biological basis."

"You mean how adaptive it is?"

"Yes. If touch didn't feel good, there'd be no species, parenthood, or survival. A mother wouldn't touch her baby in the right way unless the mother felt pleasure doing it. If we didn't like the feel of touching and patting one another, we wouldn't have had sex. Those animals who did more touching instinctively produced offspring which survived, and their genes were passed on and the tendency to touch became even stronger. We forget that touch is not only basic to our species, but the key to it."

As a fetus grows in the womb, surrounded by amniotic fluid, it feels liquid warmth, the heartbeat, the inner surf of the mother, and floats in a wonderful hammock that rocks gently as she walks. Birth must be a rude shock after such serenity, and a mother re-creates the womb comfort in various ways (swaddling, cradling, pressing the baby against the left side of her body where her heart is). Right after birth, human (and monkey) mothers hold their babies very close to their bodies. In primitive cultures, a mother keeps her baby close day and night. A baby born to one of the Pygmies of Zaire is in physical contact with someone at least 50 percent of the time, and is constantly being stroked or played with by other members of the tribe. A Kung! mother carries her baby in a *curass,* a sling that holds it upright at her side so that it can nurse, play with her bead necklaces, or interact with others. Kung! infants are in touch with others about 90 percent of the time, whereas our culture believes in exiling babies to cribs, baby carriages, or travel seats, keeping them at arm's length and out of the way.

An odd feature of touch is that it doesn't always have to be performed by another person, or even by a living thing. Maternity Hospital in Cambridge, England, discovered that if a premature baby were just placed on a lamb's-wool blanket for a day it would gain an average of fifteen grams more than usual. This was not due to additional heat from the blanket, since the ward was kept warm, but more akin to the tradition of "swaddling" infants, which increases tactile stimulation, decreases stress, and makes them feel lightly cuddled. In other experiments, snug-fitting blankets or clothes reduced the infants' heart rate, relaxed them; they slept more often in their womblike bindings.

All animals respond to being touched, stroked, poked in some way, and, in any case, life itself could not have evolved at all without touch—that is, without chemicals touching one another and form-ing liaisons. In the absence of touching and being touched, people of all ages can sicken and grow touch-starved.* In fetuses, touch is the first sense to develop, and in newborns it's automatic before the eyes open or the baby begins to make sense of the world. Soon after we're born, though we can't see or speak, we instinctively begin touching. Touch cells in the lips make nursing possible, clutch mech-anisms in the hands begin to reach out for warmth. Among other things, touch teaches us the difference between *I* and *other,* that there can be someone outside of ourselves, the mother. Mothers and infants do an enormous amount of touching. The first emotional comfort, touching and being touched by our mother, remains the ultimate memory of selfless love, which stays with us life long.

The little three-pound universe named Geoffrey, which I am stroking in long gentle caresses, has idly twisted his mouth and just as quickly untwisted it again. In other incubators around the room, other lives are stirring, other volunteers continue reaching in through portholes to help the infants begin to make sense of the world. The head research nurse of the ward, a graduate student in neonatal care, gives the Brazelton sensory test to a baby boy, who responds to a bright-red egg-rattle. Picking the baby up, she swings it gently around and its eyes go in the direction of the spin, as they should, then return to the midline. Next she rings a small schoolbell for ten seconds at each side, and repeats this four times. It is a very Buddhist scene. In a nearby crib, a preemie who is having his hearing tested wears a headset that makes him look like a telegraph operator. The policy with premature babies used to be not to disturb them any more than necessary, and they lived in a kind of isolation booth, but now the evidence about touch is so plentiful and eloquent that many hospitals encourage touching. "Did you hug your child today?" asks

---

*What a curious and deprived life the Dionne quints lived. Born in Ontario, Canada, they were seized by the government and put in a kind of zoo. So they lived in a sterile room behind bars. At one point their mother, who wasn't allowed to touch them, stood in line with the other paying viewers. Only after a lawsuit was she able to get her children back. None of them grew up normally.

the bumper sticker. As it turns out, this is more than a casual question. Touch seems to be as essential as sunlight.

WHAT IS A TOUCH?

Touch is the oldest sense, and the most urgent. If a saber-toothed tiger is touching a paw to your shoulder, you need to know right away. Any first-time touch, or change in touch (from gentle to stinging, say), sends the brain into a flurry of activity. Any continuous low-level touch becomes background. When we touch something on purpose—our lover, the fender of a new car, the tongue of a penguin—we set in motion our complex web of touch receptors, making them fire by exposing them to a sensation, changing it, exposing them to another. The brain reads the firings and stopfirings like Morse code and registers *smooth, raspy, cold.*

Touch receptors can be blanked out simply by tedium. When we put on a heavy sweater, we're acutely aware of its texture, weight, and feel against our skin, but after a while we completely ignore it. A constant consistent pressure registers at first, activating the touch receptors; then the receptors stop working. So wearing wool or a wristwatch or a necklace doesn't bother us much, unless the day heats up or the necklace breaks. When any change occurs, the receptors fire and we become suddenly aware. Research suggests that, though there are four main types of receptors, there are many others along a wide spectrum of response. After all, our palette of feelings through touch is more elaborate than just hot, cold, pain, and pressure. Many touch receptors combine to produce what we call a twinge. Consider all the varieties of pain, irritation, abrasion; all the textures of lick, pat, wipe, fondle, knead; all the prickling, bruising, tingling, brushing, scratching, banging, fumbling, kissing, nudging. Chalking your hands before you climb onto uneven parallel bars. A plunge into an icy farm pond on a summer day when the air temperature and body temperature are the same. The feel of a sweat bee delicately licking moist beads from your ankle. Reaching blindfolded into a bowl of Jell-O as part of a club initiation. Pulling a foot out of the mud. The squish of wet sand between the toes. Pressing on an angel food cake. The near-orgasmic caravan of plea- sure, shiver, pain, and relief that we call a back scratch.* On a cattle ranch some years ago, in birthing season, I helped the cowhands with the herd. Whenever we found a cow in trouble, someone had to reach into her vagina and check her condition. "You're a female," they'd invariably say, "you do it," meaning that I was bound to know, by feel, the internal landscape of another female, even if she was only distantly related to me and her organs were horizontal. "Look for the two big boulders just over a rise . . . ," a Spanish- American cowhand had said helpfully on one occasion. Up to your shoulder inside a cow, you feel the hot heavy squeeze of her, but I'll never forget my startled delight the first time I withdrew my hand slowly and felt the cow's muscles contract and release one after another, like a row of people shaking hands with me in a receiving line. I wonder if this is how it feels to be born. Also, scientists have discovered that most of the nerve receptors will respond to pressure, as well as to whatever they specialize in. For the longest time we assumed that each sensation had its own receptor and that that receptor had its own pathway to the brain, but it looks now as if the body's grasslands of neurons relate any sensation according to electri- cal codes. Pain produces irregular bleats from the nerves at jagged intervals. Itching produces a fast, regular pattern. Heat produces a crescendo as the area heats up. A little pressure produces a flurry of excitement, then fades, and a stronger pressure just extends the burst of activity.

After a while, as suggested, a touch receptor "adapts" to the stimuli and stops responding, which is just as well or we would be driven crazy by the feel of a light sweater against the skin on a cool summer's evening, or go berserk if a breeze didn't quit. This fatigue

---

*Mother tells me she once hooked a rug out of old shirts, torn underwear, and my father's socks, all slivered up like apples and plugged into burlap with crocheting tools. She must mean the black-and-floral slab that surfaced like a raft on the basement floor, ice-cold and ugly with ammonia where the stray dog we took in for the winter had worms. It's not so much the rag rug itself I have frozen in my memory as its spongy feel. After thirty years, I can still fetch back that revelation of acrylic squoosh.

doesn't happen among the deep Pacinian corpuscles and Ruffini's organs (joints) or the Golgi's organs (tendons), which give us information about our internal climate, because if they nodded we would fall down midstride. But the other receptors, so alert at first, so hungry for novelty, after a while say the electrical equivalent of "Oh, that again," and begin to doze, so we can get on with life. We may feel self-conscious much of the time, but we're not often conscious of our physical selves, or we'd be exhausted in a typhoon of sensation.

Some forms of touch irritate and delight us simultaneously. Tickling may be a combination of the signals for, say, pressure and pain. Wetness may be a mix of temperature and pressure. But when we lose touch (the dentist gives you a shot of novocaine; an arm or leg falls asleep from lowered blood supply), we feel odd and alien. Imagine how frightening it must be to lose touch permanently. Touch loss can be maddeningly specific: A person loses a sense of temperature, or of pain. When my dentist gave me a shot of Carbocaine, my jaw dropped like a slab of pottery. I could still feel pressure and temperature—though the temperature sensation was reversed (ice water tasted like water but felt hot)—but I no longer felt any level of pain in the jaw. The absence of pain's minute markers—a scratch, a pinch, a twinge—made the flesh feel cadaverous. In St. Louis, Missouri, one day a couple of years ago, I was going to a reading with novelist Stanley Elkin, who has suffered from MS for many years. Stanley could still drive, and we decided to take his car. But when we got to it and he went around to the driver's door, he stopped and stood for what seemed ages, groping in his pocket. Finally he pulled out the entire contents of the pocket and set it all on the car hood so he could *see* his keys. Many sufferers of MS can feel an object in their pocket (a set of car keys), but they can't identify it by touch. The brain won't decode the shape correctly. As those who are simultaneously deaf and blind have shown, it's possible to get on predominantly by touch, but to be without touch is to move through a blurred, deadened world, in which you could lose a leg and not know it, burn your hand without feeling, and lose track of where you stop and the featureless day begins.

THE CODE SENDERS

It takes a troupe of receptors to make the symphonic delicacy we call a caress. Between the epidermis and the dermis lie tiny egg-shaped Meissner's corpuscles, which are nerves enclosed in capsules. They seem to specialize in hairless parts of the body—the soles of the feet, fingertips (which have 9,000 per square inch), clitoris, penis, nipples, palms, and tongue—the erogenous zones and other ultrasensitive ports of call—and they respond fast to the lightest stimulation. Inside a Meissner corpuscle, like the many filaments inside a light bulb, branching, looping nerve endings lie parallel to the surface of the skin and pick up a wealth of sensation. Their parallel arrangement may make them especially sensitive to something touching them at a perpendicular angle. Furthermore, they are extremely specific because each area of the corpuscle can respond independently. As one researcher describes it, "It's as though the receptor were composed of separate coils like an innerspring mattress; one can be depressed without disturbing the others." What they record is low-frequency vibrations, the feeling of a finger stroking a beautifully woven sari, for example, or the soft angled skin inside another's elbow.

The Pacinian corpuscles respond very quickly to changes in pressure, and they tend to lie near joints, in some deep tissues, and in the genitals and mammary glands. Thick, onion-shaped sensors, they tell the brain what is pressing and also about the movement of joints or how the organs may be shifting their position when we move. It doesn't take much pressure to make them respond fast and rush messages to the brain. But they're also adept with vibrating or varying sensations, especially high-frequency ones (a violin string, for instance); indeed, it may be the onionlike layers of the corpuscle that decipher differing vibrations so well. What the Pacinian corpuscles do is convert mechanical energy into electrical energy, as Bernhard Katz of University College, London, showed in 1950 in electrical experiments with muscles. Subsequent research has led to a better understanding of this process, as Donald Carr describes in *The Forgotten Senses:*

Neurologists now believe that one can picture the touch receptor as a membrane in which there are a number of tiny holes, or at least potential holes, like a piece of Swiss cheese covered with cellophane. In the resting state the holes are too small or the cellophane too thick for certain ions to pass through. Mechanical deformation opens up these holes. When . . . currents are . . . formed . . . by a strong pressure such as a pinprick, the currents are strong enough to trigger nerve impulses and the intensity of the prick is signaled by the frequency of the impulses, since this is the only way nerve fibers can code intensity.

Our menagerie of touch receptors also includes saucer-shaped Merkel's disks, which lie just below the skin surface and respond to continuous, constant pressure (they give a sustained message, a continuous monitoring); various free nerve endings, which aren't enclosed in capsules, and respond more slowly to touch and pressure; Ruffini endings, located deep below the skin surface, which register constant pressure; temperature sensors; cylindrical heat sensors; and the most familiar, but oddest, touch receptor of all: hair.

HAIR

Hair deeply affects people, can transfigure or repulse them. Symbolic of life, hair bolts from our head. Like the earth, it can be harvested, but it will rise again. We can change its color and texture when the mood strikes us, but in time it will return to its original form, just as Nature will in time turn our precisely laid-out cities into a weedway. Giving one's lover a lock of hair to wear in a small locket* around his neck used to be a moving and tender gesture, but also a dangerous one, since to spell-casters, magicians, voodoo-ers, and necromancers of all sorts, a tuft of someone's hair could be used to cast a spell against them. In a variation on this theme, a medieval

---

*A "lock" of hair is a winding and twisting thing, according to its origin in the Indo-European *leug-*, a fascinating root at the heart of the word *locket* (in Old English "a bending together, a shutting"); as well as the Latin idea of *luxuriance, extravagance,* and *excess* (originally of plants growing in wild and unruly profusion); the Latin word for *to wrestle* (people bending around each other), as well as *to struggle* (people trying to twist and fasten events); the German word for the vegetable *leek* (because of the leaf shape); and even the Germanic word *luck* (when fate twists obliquely).

knight wore a lock of his lady's pubic hair into battle. Since one of the arch-tenets of courtly love was secrecy, choosing this tiny memento instead of a lock of hair from her head may have been more of a practical choice than a philosophical one, but it still symbolized her life-force, which he was carrying with him. Ancient male leaders wore long flowing tresses as a sign of virility (in fact, "kaiser" and "tsar" both mean "long-haired"). In the biblical story of Samson, the hero's loss of hair brings on his weakness and downfall, just as it did for the hero Gilgamesh before him. In Europe in more recent times, women who collaborated with the enemy in World War II were humiliated by having their hair cut short. Among some orthodox Jews, a young woman must cut off her hair when she marries, lest her husband find her too attractive and wish to have sex with her out of desire rather than for procreation. Rastafarians regard their dreadlocks as "high-tension cables to heaven." These days, to shock the bourgeoisie and establish their own identity, as every generation must, many young men and women wear their hair as freeform sculpture, with lacquered spikes, close-cropped patterns that resemble a formal garden maze, and colors borrowed from an aviary or spray-painted alley. The first time a student walked into my classroom wearing a "blue jay," it *did* startle me. Royal-blue slabs of hair were brushed and sprayed straight up along the sides of his head, a long jelly roll of white hair fell forward over his eyebrows, and the back was shiny black, brushed straight up and plastered close to the head. I didn't dislike it, it just seemed like a lot to fuss with each day. I'm sure my grandmother felt that way about my mother's "beehive," and I know my mother feels that way about the curly weather system which is my own mane of long thick hair. One's hairstyle can be the badge of a group, as we've always known—look at the military's crew cut, or the hairstyles worn by some nuns and monks. In the sixties, wearing long hair, especially if you were a man, often fetched a vitriolic outburst from parents, which is why the musical *Hair* summed up a generation so beautifully. The police, who seemed so clean-cut and cropped then, were succeeded by a generation of police in long sideburns and mustaches. But I remember at the Boston Love-in in 1967, my first year away at college, hearing one young man say to a passing couple who ridiculed his ponytail: "Fuck you and fuck your hairdressers." I also remember, in the fifties, walking out of my bathroom with my hair sprayed into a huge bubble. "What have you done to your hair?" my father demanded. "I've just teased it," I said. To which he replied: "Teased? You've driven it insane." I wear my curly hair au naturel these days, in a shag cut the French call *la coupe sauvage* ("the savage cut"), but its volume and faintly erotic unruliness bother my mother's sense of propriety. To her generation, serious women have serious hairdos that are formal, sprayed, and don't move. A few weeks ago, she phoned to warn me that professional women aren't taken seriously if they don't have a "wet set" (rollers, hair dryer, setting lotion, hair spray). Loose ends on one's head signal loose ends in one's life. From this point of view, which has been popular for ages, a woman grows her hair long but keeps it tightly controlled in a bun, under a hat or scarf, or with hair spray, and lets her hair down only in private at night.

Most people have about 100,000 hair follicles on their head, and lose between fifty and a hundred hairs a day through normal combing, brushing, or fussing. Each hair grows for only about two to six years, at about five or six inches a year, and then its follicle rests for a few months, the hair falls out, and is eventually replaced by a new hair. So when you see a beautiful head of hair, you're looking at hairs in many different stages in a complex system of growth, death, and renewal. Fifteen percent of it is resting at any one time, the other 85 percent growing; many dozens of hairs are all set to die tomorrow, and deep in the follicles new hairs are budding.

Hair has a tough outer coating called the cuticle, and a soft interior called the cortex. People with coarse hair have larger follicles, and also a thin outer coat (10 percent of the hair) with a large inner cortex (90 percent). People with fine hair have smaller follicles, and almost the same amount of cuticle (40 percent) as cortex (60 percent). If the follicle cells grow in an even pattern, the hair will be straight; if they grow irregularly, the hair will be curly. Lice have a hard time attaching to thick hair, which is why black schoolchildren don't succumb to epidemics of head lice as often as their white classmates. Besides being sexy to most people, head hair protects the brain from the sun's heat and ultraviolet rays, helps to insulate the skull, softens impact, and constantly monitors the world only a hair's breadth away from our body, that circle of danger and romance we allow few people to enter.

Of course, hairs grow in many places around the body, even on the toes and inside the nose and ears. The Chinese, the American Indian, and some other peoples have very little hair on their face and body; those of Mediterranean descent can be so woolly and thickly haired they seem only a step away from our ape-man ancestors. Bald men are sexy men; they go bald from a high level of testosterone in the blood, which is why you don't see bald castrati or eunuchs. Men with thick mats of hair on their shoulders and backs used to scare me. A word like "carnivore" would form in my mind when I passed them on beaches. Women tend to be smoother-fleshed than men, so it makes sense that we would shave our legs and apply lotions to accentuate the gender difference. But despite efforts to remove hair from our bodies, quite a lot remains on the arms, faces, and heads of women, and the chest, arms, and legs of men, to do what it was intended to do.

Hair is special to mammals, although reptiles do form scales, which are related. Each hair grows from the papilla, a wad of tissue at the base of a follicle, where there is a nerve ending, and there may be a group of other nerve endings nearby. The average body has about five million hairs. Because hairy skin is thinner, it's more sensitive than smooth skin. One hair can be easily triggered: If something presses it or tugs at it, if its tip is touched, if the skin around it is pressed, the hair vibrates and sparks a nerve. Down is the most sensitive hair of all and only has to move 0.00004 of an inch to make a nerve fire. Still, it can't be firing all the time, or the body would go into sensory overload. There is an infinitesimally small realm in which nothing at all seems to be happening, a desert of sensation. Then the merest breeze starts to blow, nothing like a real disturbance. When it grows just strong enough to reach an electrical threshold, it fires an impulse to the nervous system. Hairs make wonderful organs of touch. "Breeze," our brain says without much fanfare, as a few hairs on our forearms lift imperceptibly. If a dust mote or insect brushes an eyelash, we know at once and blink to protect the eye. Though hairs can take shapes as various as down or antennae, some especially useful ones are *vibrissae*—the stiff hairs cats have as whiskers—which adorn many mammals, including whales and porpoises. A cat without its whiskers bumps into things at night, and can get its head caught in tight spaces. As we can. If we ever get a say-so in evolution, one of the things I'd vote for is whiskerlike feelers to keep us from bumping into furniture, friends, or raccoons in the dark.

THE INNER CLIMATE

Some people meditate, or practice the Zen of archery. I begin each summer morning by strolling around the raised beds in my garden, where twenty-five tea and floribunda rosebushes, twenty-eight lavender and yellow day lilies, a dozen or so shade-loving plants such as hostas and monkshood, and a brilliant range of perennials and annuals flourish. It's not unusual to spend half an hour choosing a sprig of baby's breath, a pink lupine, one stem of bluebell-shaped campanella (whose stem oozes white sap—almost always a sign of poison), one orange-red rose called "Bing Crosby," one stem of red-and-white bleeding hearts, a bright yellow coriopsis, a huge fuchsia dahlia, a red-and-white, daisy-shaped miniature dahlia, and a flamboyantly speckled red-and-yellow *Pavonia tigridia* which looks like an iris that married a day lily and went to a fiesta (its name means "tiger-faced peacock," which is wonderful enough, but I've always called it a "Mexican hat dance" instead). Because I don't know what will have opened during the night or early morning, some days it's a little like discovering an emerald in your soup. Then I spend half an hour or so indoors, arranging the day's petaled baubles in a glass dish full of clear marbles, driven no doubt by laws of balance, shape, and color, but working with a calm obsessiveness that allows nothing so rude as thought to intrude.

While making a bouquet one morning, I noticed an odd thing about how we perceive temperature. Next to some cutlery soaking in a pan of hot water in the sink was one bowl of cold water and one bowl of warmish water. I put one hand in the cold, one hand in the hot. Then I put both in the warm water and, to my surprise, they gave me contradictory signals. All they were perceiving was the *movement* of temperature, not hot or cold per se. I've also noticed that, for some reason, objects of equal weight feel heavier if they're cold than if they're warm. There's no simple answer for this phenomenon. Maybe the heat receptors are more specialized, whereas the cold receptors register pressure, too.

Most of the cold receptors lie in the face, especially on the tip of the nose, the eyelids, lips, and forehead, and the genitals are sensitive to cold, as well. It's our outer shell that seems to fear cold most, acting as a sentry on perpetual watch. Receptors for warmth lie deeper in the skin, and there are fewer of them. Not surprisingly, the tongue is more sensitive to heat than many other areas of the body. If hot soup can pass the tongue test, it probably won't burn the throat and stomach. Unlike other touch information, temperature reports tell the brain of changes as well as highs and lows, and there are frequent updates. My mother used to urge me to put an ice cube on my wrist when I was too hot. This excites the cold receptors into overreacting and firing furiously. Remove the ice cube, and the wrist stays cold for quite a while. It doesn't seem like much of a poultice, but your skin only has to be warmed by three or four degrees to make you feel truly warm, only lowered by one or two degrees to make you feel decidedly cold. Then your body starts to correct things and you rub your hands together, shiver, or stick your hands under your armpits to warm up. You drink iced drinks or take a cold shower or go for a swim to cool down. On a brutally hot and humid summer day, one on which the sun feels as if it's been dipped in lye, the air is so thick it's drinkable, and your body feels like freshly melted lead, all I have to do is get into a swimming pool and stand up to my neck in cold water, ice down the brain stem, to feel rejuvenated. Why should aspirin be able to lower a fever, but not affect a normal temperature? Because it inhibits the release of the body's own pyrogen, a substance that causes fever. There are still many mysteries about the body's ability to regulate its temperature. We wake up cooler than when we go to sleep, but why should we be at our lowest temperature at about 4:00 A.M.?

Suppose we cooled the body from the inside out? In hypothermic surgery, the blood is chilled and recirculated, which reduces body temperature to about seventy-seven degrees. Science-fiction stories often involve an astronaut whose body temperature has been lowered, sleeping in suspended animation like a naked bear in a glass den. Walt Disney's family swears it isn't true, but a popular folk myth for some time now has it that Walt arranged to be frozen when he died and is lying in a magic kingdom of ice, awaiting his rebirth. Trans Time, Inc., a member of the American Cryogenics Society, does freeze people right after death, promising to bring them back to life in a later era, when the mysteries of death are scrutable and the carnage of their diseases reversible. Movies like *Ice Man* play with the idea of someone being frozen for decades or centuries, then awakening in a new world. What makes it sound so plausible, I suppose, is how familiar the scenario is in religious terms: One dies out of this life to emerge in the next. I don't think there's enough evidence that a brain and body could be frozen and defrosted without damage, but proponents of cryogenics argue that one has nothing to lose. Could there be an extreme metabolic reduction instead of freezing? The suspended animation of sci-fi stories? Different tissues have a different freezing profile, don't they? Wouldn't that mean that some would be overchilled and others underchilled? How will right-to-lifers (who are already vehemently opposed to freezing sperm, ova, and embryos) and religious zealots feel about thawing people out—what ethical debates and social turmoil will it prompt?

Warm-blooded creatures, we overheat easily and then an ancient terror sets in. We moan that we're being cooked, the way we cook other animals: "I'm roasting," we say; "I'm burning up"; "It's like an oven in here." Now that we've lost our heavy body fur we chill fast, so we must wear thick clothing when the temperature plunges. I've seen people out walking on a winter's day wearing layered clothes, wool sweaters, and bulky down coats; they look like freshly made beds on the move. The evolution of warm-blooded animals was an extraordinary breakthrough. It meant that they could keep up their body temperature despite the vagaries of the environment, and could actually migrate. Cold-blooded animals (except butterflies, eels, and sea turtles) can't migrate much, and some, like rattlesnakes and pit vipers in general, are excellent at heat detection. So are mosquitoes, moths, and other insects (which has led some researchers to conclude that people who are bitten more often than others may be radiating more heat, which makes them prime targets). Although we don't have such heat-sensing devices built into our bodies, we do create them for military use—heat-seeking missiles that strike like vipers. In recent sci-fi/horror films like *Wolfen* or *Predator*, razor-clawed, blood-lusting monsters live in a world beyond our visual range; but we are exquisitely findable by them because they can sense in infrared. The monster appears without warning, disembowels someone, and vanishes. Something about its heat-seeking ability makes it doubly horrifying. It uses one of our loveliest features to destroy us. For millennia we've relied on our warm-bloodedness as a life-force; we prize caring, compassionate people by referring to their warmth. And here is a monster homing in on that warmth. Our essence is our undoing says the message of these sensory frightmares.

Without a thick hair covering to protect us, we have to be vigilant about cold. Although the hands, feet, and other exposed parts of the body seem invaluable, since they register touch so sensitively, when cold hits they become expendable. The hands or feet can freeze, and the body still survive, but if the blood temperature lowers we're goners. So the torso responds immediately to changes in temperature, and we sense cold over a wider range of our body than we do heat. Far more women than men claim to have cold hands and feet, which isn't at all surprising. When the body gets cold, it protects the core organs first (which is why it's easy to get frostbite in your extremities); in women, it protects the reproductive organs. When your lips turn blue or your toes suffer frostbite, the blood vessels are tightening up and the body is sacrificing the extremities, sending blood to the essential inner section.

Animals love to lie in the sun and bask. Nothing looks more contented in winter than a black-and-white cocker spaniel sprawling

on the living-room carpet in a bright shaft of sunlight. Some creatures, like reptiles or houseflies, do it to regulate their body temperature, and one often sees an American alligator in a Floridian swamp
arranging itself in the sun with a voluptuary's exquisite care: one leg
and the tail under the water, the lower back and another leg in the
shadow of a bush, the head and back and front legs completely in
the sun. . . . The alligators seem quite finicky about it, but really are
governing their thermostats just as we do on a fall afternoon, when
we leave on a pullover sweater but take off our hat and gloves. The
travel industry relies on human beings' love of basking, and basking
vacations are available to most anywhere. And, though some of us
like adventure travel, most prefer to sit in the sun like a rack of
spareribs, basting ourselves regularly with sauce, and quietly frying,
taking care to turn over so we'll be cooked on both sides. Why we
love to bask isn't hard to fathom. Evolution, that haute couturier of
ingenious patterns, probably designed the sensation so that animals
would search for climates conducive to good health. But, when
enough becomes too much and an animal overheats, the skin's
smallest capillaries dilate to let the heat escape. A man's face flushes.
A rabbit's ears flush. All animals perspire in one way or another, and
the perspiration evaporates, cooling the body. *It's not the heat, it's
the humidity,* we moan on a sultry day when even a cotton shirt
seems attached by saliva to one's back. As the outside air temperature reels close to 98.6 degrees, the body starts to lose track of itself
and suffers. But if it's also humid, which means the air is saturated
with water, we still sweat to cool off in the usual way, but nothing
happens. The air's too soggy to allow sweat to evaporate. So one sits
on a porch swing in Alabama, listless and sticky, fanning oneself with
a flyer from a local construction company that, according to its
advertisement, longs to "flash your gutters," while sipping iced tea
flavored with a sprig of fresh peppermint or a leaf of pineapple sage.
On the other hand, if an animal gets too cold, most often it raises
gooseflesh and shivers—skin muscles contract (to expose a smaller
area), and the shaking that follows warms the body. Even though we
can't puff up our fur the way other animals do, either to look big and
mean or to keep warm, we have tiny leftover *erector pili* muscles that

cause some of our hairs to stand up when we're cold or scared.
Certain animals have evolved fascinating strategies for keeping
warm. Von Buddenbrock reports a German beekeeper who discovered that hives never got very cold:

> The explanation is remarkable. In the winter, tens of thousands
> of bees in a hive cluster closely together. The bees in the center
> of the cluster are warm enough when the temperature drops, but
> those in the outer layers get cold; they then begin to kick their feet
> and flap their wings rapidly—in other words, they act much as we
> do when we shiver with the cold. The main thing seems to be,
> though, that their agitation spreads through the entire cluster of
> 10,000 or more bees. The concerted efforts of the group eventually
> generates a sizable amount of heat. The temperature consequently
> rises until all the bees have calmed down, and then gradually drops
> until the same process is repeated.

Again I remember that week in December when I traveled along
the coast of California with Chris Nagano of the Los Angeles Museum's Monarch Project, finding and tagging thousands of overwintering monarch butterflies. Hanging in radiant orange garlands from
the eucalyptus trees, the butterflies would occasionally spread their
wings wide like solar collectors, or quickly flap them to warm up
before setting off to find nectar. It was easy to catch them in a net
attached to the end of a long telescoping pole, and for the most part
they just rustled quietly inside the net while we sat on the floor of
the silent, insect-free eucalyptus grove. We lifted them from the net
one at a time to check their health and sex and to see if they were
pregnant, and then glued a small postage-stamp-like tag to the top
of a wing. But some mornings it was as cool as fifty degrees, and a
monarch needs the temperature to be at least fifty-five before it can
move its flying muscles. Sometimes, when I finished tagging a butterfly and launched it into the air in the usual way—as if tossing a
hankie—it would fall right to the ground, a tasty morsel for a quick
predator. Whenever that happened, I would pick the butterfly up
by its closed wings and hold it in my open mouth while I breathed
hot air over its muscles. After a few seconds it would be warmed up

enough to fly, I would relaunch it, and it would go about its delicate business in the grove.

THE SKIN HAS EYES

Touch, by clarifying and adding to the shorthand of the eyes, teaches us that we live in a three-dimensional world. We look at a photograph taken with someone we love at a small one-llama circus in a rural town, and remember the stickiness of that summer day, the feel of the llama insinuating its velvety nose into our shirt pocket, into our hand, under our arm, and around our chest, gently but irrepressibly looking for food. At that moment, the word "llama" becomes a verb in our vocabulary, because you have to llama your way through life from time to time. We remember the feel of the loved one's hand, how his body curves, the texture of his hair. Touch allows us to find our way in the world in the darkness or in other circumstances where we can't fully use our other senses.* By combining eyesight and touch, primates excel at locating objects in space. Although there's no special name for the ability, we can touch something and decide if it's heavy, light, gaseous, soft, hard, liquid, solid. As Svetlana Alper shrewdly observes in *Rembrandt's Enterprise: The Studio and the Market* (1988), though Rembrandt often took blindness as his subject (*The Return of the Prodigal Son*, the blind Jacob, and others):

> Blindness is not invoked with reference to a higher spiritual insight, but to call attention to the activity of touch in our experience of the world. Rembrandt represents touch as the embodiment of sight. . . . And it is relevant to recall that the

---

*Touch is being used successfully as a substitute for hearing. Varying numbers of gold-plated electrodes are attached to a stimulator belt, which is usually worn on the abdomen, arm, forehead, or legs. A deaf child is taught that particular sounds have particular skin patterns. Then the teacher asks the child to create sounds that will produce the same pattern on the skin. This works especially well with words like "sue," "do," "too," "new," which are difficult for deaf people to lip-read. These "tactile vocoders," as the devices are called, can't transmit the entire speech code yet, but they can be used very effectively in conjunction with lipreading. The children using them read at levels much higher than those who just lip-read. In Dr. Kimbough Oller's program of tactile vocoder use at the University of Miami, the ultimate goal is one day to substitute the sense of touch for the sense of hearing.

analogy between sight and touch had its technical counterpart in Rembrandt's handling of paint: his exploitation of the reflection of natural light off high relief to intensify highlights and cast shadows unites the visible and the substantial.

One of the things I find thrilling about Rembrandt's portraits is all he leaves unpainted, for the eye to register but the mind to record in full. It isn't necessary to paint anything but the front brim of a boy's hat; the first dozen times you see the painting, you won't notice that all Rembrandt painted was a gesture, the merest insinuation of a hat, which the viewer's mind completes from its own experience. We have touched *round*. We know what round is when we see it. "Oh, that again, *round,*" the mind says once more, and looks for other fish to fry.

What is a sense of one's self? To a large extent, it has to do with touch, with how we feel. Our *proprioceptors* (from Latin for "one's own" receptors) keep us informed about where we are in space, if our stomachs are busy, whether or not we are defecating, where our legs, arms, head are, how we're moving, what we feel like from moment to moment. Not that our sense of self is necessarily accurate. Each of us has an exaggerated mental picture of our body, with a big head, hands, mouth, and genitals, and a small trunk; children often draw people with big heads and hands, because that is the way their body feels to them. There is so much to know at any given moment. "How are you?" a passerby asks politely in Kafka's novel *The Trial,* and the hero panics, paralyzed by the shock of being asked one question he can't possibly answer. Everyday life includes a host of similar questions, ones that aren't meant to be taken seriously but are inserted into a conversation like a quarter into the slot of a mechanical horse, and I'm often tempted to give a lengthy and prankish answer. "How are you?" a friend will ask, and I'll report straight from my proprioceptors on the state of my kidneys, nasal mucosa, blood pressure, cochlea, vaginal rugosa, digestion, and general adrenal unrest. Touch fills our memory with a detailed key as to how we're shaped. A mirror would mean nothing without touch. We are forever taking the measure of ourselves in unconscious ways—idly running one hand along a forearm, seeing if our thumb and forefinger can bracelet our wrist or if we can touch our tongue to our nose or bend our thumb all the way back, feeling the length of our leg as we "ladder" our nylon from the ankle to the thigh, nervously twisting a strand of hair. But, above all, touch teaches us that life has depth and contour; it makes our sense of the world and ourself three-dimensional. Without that intricate feel for life there would be no artists, whose cunning is to make sensory and emotional maps, and no surgeons, who dive through the body with their fingers.

ADVENTURES IN THE TOUCH DOME

Going out to San Francisco, I open a not-to-be-opened-till-in-flight present from a friend—an exquisite blue-and-gold silk brocade box, inside of which lie two mirror-perfect chrome balls, each in its own silken socket. They bring to mind the mad Captain Queeg, who obsessively rotated two ball bearings while he spoke of pilfered strawberries. Inside the lid, a folded note explains:

> Ancient mandarins dating back 800 years believed these Chinese Exercise Balls induced well-being of the body and serenity of spirit. These treasured gifts were given to President Reagan and his wife while visiting the People's Republic of China. The Chinese say that rotating the balls in the palm of each hand stimulates the fingers and acupuncture points, and improves circulation of vital energy throughout the body. Sports enthusiasts, musicians, computer users and health-conscious people everywhere consider them great muscle conditioners. Arthritis sufferers feel a decided benefit from this gentle but challenging exercise. Very effective for relaxation and meditation, Chinese Exercise Balls emit a distantly mysterious chime as you turn them. Beautifully handcrafted, 45mm. hollow polished chrome balls are perfectly weighted and fit comfortably into the average man's or woman's hand.

Lifting them out one at a time, I marvel at their smoothness and slither, the *ping* they make colliding, and how relaxing it is to fidget them round and round, world over gleaming world in my hand. Actually, they look like *rin no tan*, specially weighted Oriental pleasure balls that a woman may insert into her vagina; when she rocks back and forth, the balls moving inside her give her the thrusting feeling of intercourse.

Though a trifle arcane, this is a fitting gift for a trip to San Francisco's Touch Dome, at the door of which I arrive a few hours later. At the far end of the Exploratorium, an extraordinary hands-on science museum, stands a three-dimensional maze through which one walks, climbs, crawls, and slithers in marmoreal darkness. The pliant walls give birth to you, or fall away to a sloping floor, or guide you to a sea of what feels like navy beans, or leave you grasping your way forward among rope hammocks. Now and then your hand strays over a familiar shape—a brush, a sandal—which seems as startling as a flash flood, and then you return to the indecipherable dark again. A few people get violently claustrophobic and start screaming, and then a guard sneaks in to rescue them, but even people who aren't normally claustrophobic have moments of sheer panic when they wonder if they will indeed find their way back to the world of sight. The blackness is as perfect as solid rock, and the maze tumbles into slides too narrow to sit up in. You can feel the beginning of the slide and its rough dimensions, but not its length or how it might change farther on. How far will it plunge? Suppose you get trapped midway, unable to lift your head or move your arms? If you go arms-first, to feel your way along, suppose it narrows and you are unable to back up? Suppose there is a drop at the bottom into a soft surface, which you will enter headfirst? Down you slide, hands over head, somersaulting free a few moments later. Crawling into a room that seems to have no outlet, you stretch high and discover handholds, then climb blindly up and pull yourself to another level of maze. Something light and sticky brushes your face, the blackness becomes a solid mystery again, disorienting and full of blind alleys; the darkness pours its panic marbles under your feet and you stumble at speed into a quagmire of something dry but mobile that surges up to your knees; then, heart pounding, you trip through thick rubber fringes, grab hold, and fall down a ramp into bright light, having survived a small expedition of pure touch.

ANIMALS

Human beings may be voluptuaries of touch, but animals are the real touch masters. Sponges have a profound sense of touch; they feel every quiver in the water. Tapeworms are thought to use only touch to perceive the world. Insect-eating plants live mainly by touch. Cockroaches have paved appendages, called *cerci,* at the base of the abdomen, which are so responsive to vibration that the insects are frequently used in laboratory experiments related to touch. Snails have extremely sensitive feet. Alligators and crocodiles use the many touch receptors around their heads to engage in elaborate stroking and necking during courtship. Though one imagines a turtle's shell to be without feeling, large sea turtles enjoy having their shells lightly scratched, and they can feel an object as delicate as a twig moving across it. Any animal that digs for a living, such as a prairie dog or an anteater, or must live by night, usually has a great sense of touch. The Eimer organ (a Pacinian-like corpuscle in the snout of a mole) can sense the slightest disturbances in the soil that might mean the presence of an earthworm nearby. The duck's bill is very sensitive to water vibrations because its skin contains Herbst corpuscles similar to Pacinian corpuscles. A woodpecker uses its tongue—which also contains a Herbst corpuscle—to search for insects in the wood it has thrilled. Penguins must touch to survive—they stand on their parents' feet and press close to their warm bellies—and so have developed a real passion for touching and being touched. Rats are compulsive touchers. Some aquatic animals can feel vibrations in the water over large distances, and detect with great precision anything moving in their vicinity. Touch is a powerfully important sense among animals, for whom the slightest touch of an object or another animal triggers responses. One need only watch the body whims of a house cat rubbing and wrapping itself around its owner's leg, or the courting of two giraffes thwacking their long necks together. And many animals enjoy touch games for hours on end, whether it is two dogs, their tongues flopping, playing chase and tumble on the lawn, or a pack of teenage boys playing "touch" football in a corner lot.

Folk wisdom has it that animals can predict earthquakes. Livestock are often reported busting out of their barns, household pets leaping from the house, pacing in a frenzy, or simply acting strangely before a tremor, which may be because of the static electricity in the air. As Helmut Tributsch, of the Free University of Berlin, realized, an animal's skin is much drier than that of a human being. There is a lot of electromagnetic upset just before an earthquake, and this produces static electricity, which makes an animal's hair stand up and quiver. I remember watching the launch of Viking II at Cape Canaveral in 1975, and how, during lift-off, the air felt itchy and electric. I felt bristlingly alert, because it was the first time in the history of our planet that we were launching a spacecraft to search for life elsewhere, and the sense of vigil deeply moved me. The launch itself produced an electromagnetic upset much like that of an earthquake and increased the static electricity in the air, which made my flesh creep. Even those skeptics among us viewers could not have been left unmoved, what with the hair standing up on their necks, the shock waves pounding on their chests like giant fists, their minds alert from the stimulating dance of negative ions, and the distant spacecraft lurching upward on spasms of apricot fire.

TATTOOS

Of all the skin-deforming arts, one of the most interesting and ancient is tattoo, which traveled like gossip over trade routes and continents. Neolithic farmers tattooed their faces with a design of blue tridents; female singers, dancers, and prostitutes in ancient Egypt wore tattoos. In 1769, Captain Cook reported in his journal that both the men and women of Tahiti displayed tattoos (a word that probably comes from the Tahitian *tatau*, "to strike"). King George V, Nicholas II, and Lady Randolph Churchill all had tattoos, along with souvenir-crazed Americans and fashionable Victorian women who wished a permanent pink to their lips. The Maori of New Zealand perfected an especially intricate style of tattoo, which Terry Landau reports on in *About Faces:*

> [They have] an elaborate tattoo technique called *moko*. . . . One traveler described a tribal chief who prided himself on having spared no visible part of his body: even his lips, tongue, gums, and palate were completely tattooed.

Japanese tattoo, called *irezumi*, is a serious folk art like landscape painting or flower arranging, and great tattoo masters still perform their Chagall-like work in full-body tattoos that are subtle, repulsive, magical, seductive, sensuous, three-dimensional, thought-provoking, and macabre.

Ultimately, tattoos make unique the surface of one's self, embody one's secret dreams, adorn with magic emblems the Altamira of the flesh. It is also a form of self-destruction; fully tattooed people live shorter lives because their skin can't breathe properly and some of the inks are poisonous. Those with tattooed faces, hands, and heads have chosen, in a way, to seal themselves off from normal society forever, and so it is not surprising that the largest number of the tattooed in Japan belong to the underworld. Tattoo masters often help the Tokyo police identify bodies. A person completely tattooed in a single coherent scene dictated by body contour and self-image makes you wonder about symbolism, decoration, and identity. In her book of forty-six almost life-size Polaroid reproductions, *The Japanese Tattoo,* photographer Sandi Fellman explains her attraction to tattoos as an infatuation with paradox: "Beauty created through brutal means," "power bestowed at the price of submission," "the glorification of the flesh as a means to spirituality."

Just as westerners donate their organs after death, a Japanese wearing the work of a grand tattoo master may donate his skin to a museum or university. Tokyo University has three hundred such masterpieces, framed. To walk into this chamber of skins must fill one with shock and wonder: What a marvel to see so many lives at full stretch, defined by needles and ink, so many people who wished to become their own text.

PAIN

In the sand-swept sprawl of the panoramic film *Lawrence of Arabia* a scene of quintessential machismo stands out: T. E. Lawrence holding his hand over a candle flame until the flesh starts to sizzle. When his companion tries the same thing, he recoils in pain, crying "Doesn't that hurt you?" as he nurses his burned hand. "Yes," Lawrence replies coolly. "Then what is the trick?" the companion asks. "The trick," Lawrence answers, "is not to mind."

One of the great riddles of biology is why the experience of pain is so subjective. Being able to withstand pain depends to a considerable extent on culture and tradition. Many soldiers have denied pain despite appalling wounds, not even requesting morphine, although in peacetime they would have demanded it. Most people going into the hospital for an operation focus completely on their pain and suffering, whereas soldiers or saints and other martyrs can think about something nobler and more important to them, and this clouds their sense of pain. Religions have always encouraged their martyrs to experience pain in order to purify the spirit. We come into this world with only the slender word "I," and giving it up in a sacred delirium is the painful ecstasy religions demand. When a fakir runs across hot coals, his skin does begin to singe—you can smell burning flesh; he just doesn't feel it. In Bali a few years ago, my mother saw men go into trances and pick up red-hot cannonballs from an open fire, then carry them down the road. As meditation techniques and biofeedback have shown, the mind can learn to conquer pain. This is particularly true in moments of crisis or exaltation, when concentrating on something outside oneself seems to distract the mind from the body, and the body from suffering and time. Of course, there are those who welcome pain in order to surmount it. In 1989, I read about a new craze in California: well-to-do business people taking weekend classes in hot-coal-walking. Pushing the body to or beyond its limits has always appealed to human beings. There is a part of our psyche that is pure timekeeper and weather watcher. Not only do we long to know how fast we can run, how high we can jump, how long we can hold our breath under water—we also like to keep checking these limits regularly to see if they've changed. Why? What difference does it make? The human body is miraculous and beautiful, whether it can "clean and jerk" three hundred pounds, swim the English Channel, or survive a year riding the subway. In anthropological terms, we've come to be who we are by evolving sharper ways to adapt to the environment, and, from the outset, what has guided us has been an elaborate system of rewards. Small wonder we're addicted to quiz shows and lotteries, paychecks and bonuses. We've always explored our mental limits, too, and pushed them without letup. In the early eighties, I spent a year as a soccer journalist, following the dazzling legwork of Pelé, Franz Beckenbauer, and virtually every other legendary international star the New York Cosmos had signed up for equally legendary sums of American cash. Choose your favorite sport; now imagine seeing all the world's best players on one team. I was interested in the ceremonial violence of sports, the psychology of games, the charmed circle of the field, the breezy rhetoric of the legs, the anthropological spectacle of watching twenty-two barely clad men run on grass in the sunlight, hazing the quarry of a ball toward the net. The fluency and grace of soccer appealed for a number of reasons, and I wanted to absorb some of its atmosphere for a novel I was writing. I was amazed to discover that the players frequently realized only at halftime or after a match that they'd hurt themselves badly and were indeed in wicked pain. During the match, there hadn't been the rumor of pain, but once the match was over and they could afford the luxury of suffering, pain screamed like a noon factory whistle.

Often our fear of pain contributes to it. Our culture expects childbirth to be a deeply painful event, and so, for us, it is. Women from other cultures stop their work in the fields to give birth, returning to the fields immediately afterward. Initiation and adolescence rites around the world often involve penetrating pain, which initiates must endure to prove themselves worthy. In the sun dance of the Sioux, for instance, a young warrior would allow the skin of his chest to be pierced by iron rods; then he was hung from a stanchion. When I was in Istanbul in the 1970s, I saw teenage boys dressed in shiny silk fezzes and silk suits decorated with glitter. They were preparing for circumcision, a festive event in the life of a Turk, which occurs at around the age of fifteen. No anesthetic is used; instead, a boy is given a jelly candy to chew. Sir Richard Burton's writings abound with descriptions of tribal mutilation and torture rituals, including one in which a shaman removes an apron of flesh from the front of a boy, cutting all the way from the stomach to the thighs, producing a huge white scar.

Women in some cultures go through many painful initiation rites, often including circumcision, which removes or destroys the clitoris. Being able to endure the pain of childbirth is expected of women, but there are also disguised rites of pain, pain that is endured for the sake of health or beauty. Women have their legs waxed as a matter of fashion, and have done so throughout the ages. When mine were waxed at a Manhattan beauty salon recently, the pain, which began like 10,000 bees stinging me simultaneously, was excruciating. Change the woman from a Rumanian cosmetician to a German Gestapo agent. Change the room from a cubicle in a beauty emporium to a prison cell. Keep the level of pain exactly the same, and it easily qualifies as torture. We tend to think of torture in the name of beauty as an aberration of the ancients, but there are modern scourging parlors. People have always mutilated their skins, often enduring pain to be beautiful, as if the pain chastened the beauty, gave it the special veneer of sacrifice. Many women experience extreme pain during their periods each month, but they accept the pain because they understand that it's not caused by someone else, it's not malicious, and it doesn't surprise them; and this makes all the difference.

There are also illusions of pain as vivid as optical illusions, times when the sufferer imagines he or she feels pain that cannot possibly exist. In some cultures, the father experiences a false pregnancy— *couvade* as it's called—and takes to bed with childbirth pains, going through his own arduous experience of having a baby. The internal organs don't have many pain receptors (the skin is supposed to be the guard post), so people often feel "referred pain" when one of their organs is in trouble. Heart attacks frequently produce a pain in the stomach, the left arm, or the shoulder. When this happens, the brain can't figure out exactly where the message is coming from. In the classic phenomenon of phantom-limb pain, the brain gets faulty signals and continues to feel pain in a limb that has been amputated; such pain can be torturous, perverse, and maddening, since there is nothing physically present to hurt.

Pain has plagued us throughout the history of our species. We spend our lives trying to avoid it, and, from one point of view, what we call "happiness" may be just the absence of pain. Yet it is difficult to define pain, which may be sharp, dull, shooting, throbbing, imaginary, or referred. We have many pains that surge from within as cramps and aches. And we also talk about emotional distress as pain. Pains are often combined, the emotional with the physical, and the physical with the physical. When you burn yourself, the skin swells and blisters, and when the blister breaks, the skin hurts in yet another way. A wound may become infected. Then histamine and serotonin are released, which dilate the blood vessels and trigger a pain response. Not all internal injuries can be felt (it's possible to do brain surgery under a local anesthetic), but illnesses that constrict blood flow often are: Angina pectoris, for example, which occurs when the coronary arteries shrink too tight for blood to comfortably pass. Even intense pain often eludes accurate description, as Virginia Woolf reminds us in her essay "On Being Ill": "English, which can express the thoughts of Hamlet and the tragedy of Lear, has no words for the shiver and the headache . . . let a sufferer try to describe a pain in his head to a doctor and language at once runs dry."

EASING PAIN

Just as there are many forms of pain, there are many remedies for it. Anesthetics like novocaine or cocaine either block the body's ability to send high-frequency pain signals to the brain or will not allow sodium to flow into the nerve cell. Some drugs manage to confuse the signals given at different stages of the pain message. Naturally occurring opiates called endorphins occupy the receptor sites so that they can't receive the neural transmitter's message of

pain.* Cocaine interferes with the neural transmitters in just this
way. Part of the reason heroin addicts need more and more of the
drug to get high is because that drug causes the body to produce less
of its own endorphins and begins to depend on the heroin to take
over their task. This increased threshold can also happen among
arthritis sufferers or other long-term heavy users of simple analgesics.
Aspirin works by inhibiting the flow of substances that stimulates
pain receptors when you have an injury, so that you don't receive as
many pain impulses. Continuous use of any analgesic can neutralize
its beneficial effect, but only twenty minutes of aerobic exercise is
enough to stimulate the body to produce more endorphins, natural
painkillers. Shifting your attention to something else will distract you
from pain; pain requires our full attention. A simple and effective
form of pain relief comes from "lateral inhibition": If a mob of
neurons all try to respond at once they get blocked. If you stub your
toe and then rub the area around it, the pain will subside in the mass
confusion. If you apply ice to a bruise, it will not only help with
swelling, it will also transmit cold messages instead of pain messages.
During sex, we tend not to mind a certain amount of pain (indeed,
for some people, pain seems to heighten the pleasure) and that may
be because of lateral stimulation—the brain is receiving so many
pleasure signals it doesn't pay much attention to those of moderate
pain. Relaxation techniques, hypnosis, acupuncture, and placebos
can fool the body into producing endorphins, and stop the pain
message from being sent out. We don't feel electricity, of course,
we feel sensations; but if the electrical code for pain isn't handed
around, we don't feel pain. Human beings can withstand enormous
amounts of pain (women have higher pain thresholds than men), but
not without chemical help, or sleight of mind. During pregnancy,
endorphin levels rise as the time of delivery gets closer. One researcher has even suggested that a pregnant woman craves certain

---

*The Ebers papyrus, a sixteenth-century Egyptian medical handbook, refers to opium as a
painkiller. The ancients understood that opiates dulled pain, but it was only recently that
people began to understand how. In the fifth century, Hippocrates was using willow bark, from
which aspirin is derived.

foods because they're high in substances that produce serotonin, which the woman will need to endure the pain of childbirth.

I once knew a songwriter with a lovely sherbety voice, who played guitar and sang in nightclubs in Pennsylvania. At the age of twenty-eight her arthritis was so acute that she had to loosen up her hands before each performance by baking them in gloves of warm wax. In time the pain grew too stubborn, and she gave up performing for teaching. For long-term sufferers, "Pain is greedy, boorish, meanly debilitating," as neurologist Russell Martin says in *Matters Gray and White*. "It is cruel and calamitous and often constant, and, as its Latin root *poena* implies, it is the corporeal *punishment* each of us ultimately suffers for being alive." In a number of specialized pain-control centers around the country, it's understood that pain is as much an emotional and psychological affliction as a physical one. Teams of neurologists, psychologists, physical therapists, and other angologists (people who study pain) work with those disabled by chronic pain, and try to find ways through the madness of their patients' bodies.

THE POINT OF PAIN

Why human beings feel pain has been the subject of theological debate, philosophical schisms, psychoanalytical edicts, and mumbo jumbo for centuries. Pain was the punishment for wrongdoing in the Garden of Eden. Pain was the price one paid for not being morally perfect. Pain was a self-affliction brought about by sexual repression. Pain was dished out by vengeful gods, or was the result of falling out of harmony with nature. Indeed, our word *holy* goes back through Old English to *haelan*, "to heal," and the Indo-European *kailo*, which meant "whole" or "uninjured." The purpose of pain is to warn the body about possible injury. Millions of free nerve endings alarm us; whenever they're hit, we feel pain. Slam our elbow against a bookcase, and, as Russell Martin describes the process:

. . . a number of chemical substances such as prostaglandins, histamine, bradykinin, and others stored in or near the nerve endings at the site of the injury are suddenly released. Prostaglandins quickly increase blood circulation to the damaged area, facilitating the infection-fighting and healing functions of the blood's white cells, antibodies, and oxygen. Together with bradykinin and other substances, present in only minute quantities, prostaglandins also stimulate the nerve endings, causing them to transmit electrical impulses along the length of the affected sensory nerve to its junction with the "dorsal horn" of the spinal cord, a strip of gray-matter tissue running the length of the spinal cord, which collects sensory signals from all parts of the body and relays them to the brain—first to the thalamus, where pain is first "felt," then on to the "sensory strip" of the cerebral cortex, where the pain becomes conscious, its location and intensity perceived.

According to the *pattern theory,* nerve impulses combine to telegraph those Morse-code-like messages of pain. Some pains just rush to the spinal cord, so that we can flinch if we touch a hot stove; and we call this a reflex, by which we mean that, as we always suspected, we can act without thinking and we frequently do. Acute pain—a ripped ligament, a burn—hurts so badly that we'll immobilize part of the body long enough for it to heal. A prick of the skin may not hurt the most, but it hurts the fastest, the signal traveling to the brain at ninety-eight feet per second. Burning or aching travels slower (about six and a half feet per second). Leg pains sometimes travel at up to 290 miles per hour. We pay no attention to our internal workings unless something goes wrong, when we might feel hunger pangs, or headaches, or thirst. Still, scientists do not agree on exactly what pain is. Some say it's a response of specific receptors to specific dangers—noxious chemicals, burning, stabbing or cutting, freezing—and others feel that it's much more ambiguous, an extreme sensory stimulation of any kind, because, in the delicate ecosystem of our body, too much of anything will disturb the balance. So, in this sense, pain really is a sign that we're out of harmony with Nature. When we're in pain the localized place hurts but the entire body responds. We grow sweaty, our pupils dilate, our blood pressure shoots up. Oddly enough, the same thing happens when we're angry or scared. There is a deep emotional component to pain. If we're badly hurt, we might also be afraid. And what are we to make of those individuals who are sadomasochists, who combine pleasure with pain?

In his famous experiments, Ivan Pavlov gave dogs a strong electric shock, which pained them severely. Then he fed them each day after a painful shock, conditioning them to associate the shock with something positive. Even when he increased the strength of the shock, they wagged their tails and salivated in expectation of dinner. In other experiments, he allowed cats to hit a switch that shocked them and fed them at the same time, and found they were eager to put up with the shock in order to get the food.

Kafka wrote short stories in which people endure pain professionally, as "hunger artists" or other self-mutilators; audiences often pay for the dubious privilege of watching someone suffer. There have always been performers of pain, artists of self-mutilation, to whom pain has a different meaning than for the rest of us. Edward Gibson, a turn-of-the-century vaudeville performer billed as "the human pincushion," let customers stick pins into him and at one point acted out a crucifixion on stage, nails piercing his hands and feet. It was only because people in the audience started fainting that authorities stopped his performance. Then there was the notorious German self-mutilator, Rudolf Schwarzkogler, whose "performances" of self-inflicted razor slashes and knife wounds filled a public hungry for sadism with unparallelled horror. Do these people not feel pain at all? Are their pleasure and pain centers cross-wired by mistake? Or, like T. E. Lawrence, do they feel pain in all its molten terror and not mind?

KISSING

Sex is the ultimate intimacy, the ultimate touching when, like two paramecia, we engulf one another. We play at devouring each other, digesting each other, we nurse on each other, drink each other's fluids, actually get under each other's skin. Kissing, we share one breath, open the sealed fortress of our body to our lover. We shelter under a warm net of kisses. We drink from the well of each other's mouths. Setting out on a kiss caravan of the other's body, we map the new terrain with our fingertips and lips, pausing at the oasis of a nipple, the hillock of a thigh, the backbone's meandering riverbed. It is a kind of pilgrimage of touch, which leads us to the temple of our desire.

We most often touch a lover's genitals before we actually see them. For the most part, our leftover puritanism doesn't condone exhibiting ourselves to each other naked before we've kissed and fondled first. There is an etiquette, a protocol, even in impetuous, runaway sex. But kissing can happen right away, and, if people care for each other, then it's less a prelude to mating than a sign of deep regard. There are wild, hungry kisses or there are rollicking kisses, and there are kisses fluttery and soft as the feathers of cockatoos. It's as if, in the complex language of love, there were a word that could only be spoken by lips when lips touch, a silent contract sealed with a kiss. One style of sex can be bare bones, fundamental and unromantic, but a kiss is the height of voluptuousness, an expense of time and an expanse of spirit in the sweet toil of romance, when one's bones quiver, anticipation rockets, but gratification is kept at bay on purpose, in exquisite torment, to build to a succulent crescendo of emotion and passion.

When I was in high school in the early sixties, nice girls didn't go all the way—most of us wouldn't have known how to. But man, could we kiss! We kissed for hours in the busted-up front seat of a borrowed Chevy, which, in motion, sounded like a broken dinette set; we kissed inventively, clutching our boyfriends from behind as we straddled motorcycles, whose vibrations turned our hips to jelly; we kissed extravagantly beside a turtlearium in the park, or at the local rose garden or zoo; we kissed delicately, in waves of sipping and puckering; we kissed torridly, with tongues like hot pokers; we kissed timelessly, because lovers throughout the ages knew our longing; we kissed wildly, almost painfully, with tough, soul-stealing rigor; we kissed elaborately, as if we were inventing kisses for the first time; we kissed furtively when we met in the hallways between classes; we kissed soulfully in the shadows at concerts, the way we thought musical knights of passion like The Righteous Brothers and their ladies did; we kissed articles of clothing or objects belonging to our boyfriends; we kissed our hands when we blew our boyfriends kisses across the street; we kissed our pillows at night, pretending they were mates; we kissed shamelessly, with all the robust sappiness of youth; we kissed as if kissing could save us from ourselves.

Just before I went off to summer camp, which is what fourteen-year-old girls in suburban Pennsylvania did to mark time, my boyfriend, whom my parents did not approve of (wrong religion) and had forbidden me to see, used to walk five miles across town each evening, and climb in through my bedroom window just to kiss me. These were not open-mouthed "French" kisses, which we didn't know about, and they weren't accompanied by groping. They were just earth-stopping, soulful, on-the-ledge-of-adolescence kissing, when you press your lips together and yearn so hard you feel faint. We wrote letters while I was away, but when school started again in the fall the affair seemed to fade of its own accord. I still remember those summer nights, how my boyfriend would hide in my closet if my parents or brother chanced in, and then kiss me for an hour or so and head back home before dark, and I marvel at his determination and the power of a kiss.

A kiss seems the smallest movement of the lips, yet it can capture emotions wild as kindling, or be a contract, or dash a mystery. Some cultures just don't do much kissing. In *The Kiss and Its History*, Dr. Christopher Nyrop refers to Finnish tribes "which bathe together in a state of complete nudity," but regard kissing "as something indecent." Certain African tribes, whose lips are decorated, mutilated, stretched or in other ways deformed, don't kiss. But they are unusual. Most people on the planet greet one another face to face; their greeting may take many forms, but it usually includes kissing, nose-kissing, or nose-saluting. There are many theories about how kissing began. Some authorities, as noted, believe it evolved from the act of smelling someone's face, inhaling them out of friendship or love in order to gauge their mood and well-being. There are cultures today in which people greet one another by putting their heads together and inhaling the other's essence. Some sniff each other's hands. The mucous membranes of the lips are exquisitely sensitive, and we often use the mouth to taste texture while using the nose to smell flavor. Animals frequently lick their masters or their young with relish, savoring the taste of a favorite's identity.* So we may indeed have begun kissing as a way to taste-and-smell someone. According to the Bible account, when Isaac grew old and lost his sight, he called his son Esau to kiss him and receive a blessing, but Jacob put on Esau's clothing and, because he smelled like Esau to his blind father, received the kiss instead. In Mongolia, a father does not kiss his son; he smells his son's head. Some cultures prefer just to rub noses (Inuits, Maoris, Polynesians, and others), while in some Malay tribes the word for "smell" means the same as "salute." Here is how Charles Darwin describes the Malay nose-rubbing kiss: "The women squatted with their faces upturned; my attendants stood leaning over theirs, and commenced rubbing. It lasted somewhat longer than a hearty handshake with us. During this process they uttered a grunt of satisfaction."

Some cultures kiss chastely, some kiss extravagantly, and some kiss more savagely, biting and sucking each other's lips. In *The Customs of the Swahili People,* edited by J. W. T. Allen, it is reported that a Swahili husband and wife kiss on the lips if they are indoors, and will freely kiss young children. However, boys over the age of seven usually are not kissed by mother, aunt, sister-in-law, or sister. The father may kiss a son, but a brother or father shouldn't kiss a girl. Furthermore,

> When his grandmother or his aunt or another woman comes, a child one or two years old is told to show his love for his aunt, and he goes to her. Then she tells him to kiss her, and he does so. Then he is told by his mother to show his aunt his tobacco,

---

*Not only humans kiss. Apes and chimps have been observed kissing and embracing as a form of peacemaking.

and he lifts his clothes and shows her his penis. She tweaks the penis and sniffs and sneezes and says: "O, very strong tobacco." Then she says, "Hide your tobacco." If there are four or five women, they all sniff and are pleased and laugh a lot.

How did mouth-kissing begin? To primitive peoples, the hot air wafting from their mouths may have seemed a magical embodiment of the soul, and a kiss a way to fuse two souls. Desmond Morris, who has been observing people with a keen zoologist's eye for quite a while, is one of a number of authorities who claim this fascinating and, to me, plausible origin for French kissing:

> In early human societies, before commercial baby-food was invented, mothers weaned their children by chewing up their food and then passing it into the infantile mouth by lip-to-lip contact— which naturally involved a considerable amount of tonguing and mutual mouth-pressure. This almost bird-like system of parental care seems strange and alien to us today, but our species probably practiced it for a million years or more, and adult erotic kissing today is almost certainly a Relic Gesture stemming from these origins. . . . Whether it has been handed down to us from generation to generation . . . or whether we have an inborn predisposition towards it, we cannot say. But, whichever is the case, it looks rather as though, with the deep kissing and tonguing of modern lovers, we are back again at the infantile mouth-feeding stage of the far-distant past. . . . If the young lovers exploring each other's mouths with their tongues feel the ancient comfort of parental mouth-feeding, this may help them to increase their mutual trust and thereby their pair-bonding.

Our lips are deliciously soft and responsive. Their touch sensations are represented by a large part of the brain, and what a boon that is to kissing. We don't just kiss romantically, of course; we also kiss dice before we roll them, kiss our own hurt finger or that of a loved one, kiss a religious symbol or statue, kiss the flag of our homeland or the ground itself, kiss a good-luck charm, kiss a photograph, kiss the king's or bishop's ring, kiss our own fingers to signal farewell to someone. The ancient Romans used to deliver the "last kiss," which custom had it would capture a dying person's soul.* In America, we "kiss off" someone when we dump them, and they yell "Kiss my ass!" when angry. Young women press lipsticked mouths to the backs of envelopes so all the tiny lines will carry like fingerprint kisses to their sweethearts. We even refer to billiard balls as "kissing" when they touch delicately and glance away. Hershey sells small foil-wrapped candy "kisses," so we can give love to ourselves or others with each morsel. Christian worship includes a "kiss of peace," whether of a holy object—a relic or a cross—or of fellow worshippers, translated by some Christians into a rather more restrained handshake. William S. Walsh's 1897 book, *Curiosities of Popular Customs*, quotes a Dean Stanley, writing in *Christian Institutions*, as reporting travelers who "have had their faces stroked and been kissed by the Coptic priest in the cathedral at Cairo, while at the same moment everybody else was kissing everybody throughout the church." In ancient Egypt, the Orient, Rome, and Greece, honor used to dictate kissing the hem or feet or hands of important persons. Mary Magdalen kissed the feet of Jesus. A sultan often required subjects of varying ranks to kiss varying parts of his royal body: High officials might kiss the toe, others merely the fringe of his scarf. The riffraff just bowed to the ground. Drawing a row of XXXXXs at the bottom of a letter to represent kisses began in the Middle Ages, when so many people were illiterate that a cross was acceptable as a signature on a legal document. The cross did not represent the Crucifixion, nor was it an arbitrary scrawl; it stood for "St. Andrew's mark," and people vowed to be honest in his sacred name. To pledge their sincerity, they would kiss their signature. In time, the "X" became associated with the kiss alone.†

Perhaps the most famous kiss in the world is Rodin's sculpture *The Kiss*, in which two lovers, sitting on a rocky ledge or outcrop-

---

*Last-kiss scenes appear in Ovid's *Metamorphoses* (VIII, 860–61), Seneca's *Hercules Oetaeus*, and Virgil's *Aeneid* (IV, 684–85), among others, and in a more erotic form in the writings of Ariosto.

†It used to be fashionable in Spain to close formal letters with QBSP (*Que Besa Su Pies*, "Who kisses your feet") or QBSM (*Que Besa Su Mano*, "Who kisses your hand").

ping, embrace tenderly with radiant energy, and kiss forever. Her left hand wrapped around his neck, she seems almost to be swooning, or to be singing into his mouth. As he rests his open right hand on her thigh, a thigh he knows well and adores, he seems to be ready to play her leg as if it were a musical instrument. Enveloped in each other, glued together by touch at the shoulder, hand, leg, hip, and chest, they seal their fate and close it with the stoppers of their mouths. His calves and knees are beautiful, her ankles are strong and firmly feminine, and her buttocks, waist, and breasts are all heavily fleshed and curvy. Ecstasy pours off every inch of them. Touching in only a few places, they seem to be touching in every cell. Above all, they are oblivious to us, the sculptor, or anything on earth outside of themselves. It is as if they have fallen down the well of each other; they are not only self-absorbed, but actually absorbing one another. Rodin, who often took secret sketch notes of the irrelevant motions made by his models, has given these lovers a vitality and thrill that bronze can rarely capture in its fundamental calm. Only the fluent, abstracted stroking and pressing of live lovers actually kissing could capture it. Rilke notes how Rodin was able to fill his sculptures "with this deep inner vitality, with the rich and amazing restlessness of life. Even the tranquility, where there was tranquility, was composed of hundreds upon hundreds of moments of motion keeping each other in equilibrium. . . . Here was desire immeasurable, thirst so great that all the waters of the world dried in it like a single drop."

According to anthropologists, the lips remind us of the labia, because they flush red and swell when they're aroused, which is the conscious or subconscious reason women have always made them look even redder with lipstick. Today the bee-stung look is popular; models draw even larger and more hospitable lips, almost always in shades of pink and red, and then apply a further gloss to make them look shiny and moist. So, anthropologically at least, a kiss on the mouth, especially with all the plunging of tongues and the exchanging of saliva, is another form of intercourse, and it's not surprising that it should make the mind and body surge with gorgeous sensations.

THE HAND

1988: Summer in upstate New York passes in a slow, humid embrace. The big event this week is a convention of psychics meeting at the Ramada Inn downtown, to tell fortunes and swap stories. Classes and special events take place in nearby rooms, but for a small fee the general public can enter the main ballroom, and choose to visit one of the many booths arcing around the walls in a horseshoe, or browse through the parapsychology books laid out on bridge tables in the center of the room. There are palm-readers, numerologists, telekinesis and UFO specialists, as well as men and women perched over crystal balls and Tarot cards. One tall thin woman wearing a tie-dyed muumuu works at a large easel with pastels. Not only does she do "past-life regressions," she draws the incarnations, complete with "past-life guides," as she talks about them. Watching for a while at a polite distance, I notice that many of the local people seem to have Indian guides whose names consist mainly of consonants.

Finally I decide on a palm-reader with a serious face and a bouffant, country-and-western hairdo, whose literature recounts her cavalcade of solved crimes and timely predictions. Giving her husband-manager twenty-five dollars for a short reading, I sit down across from her at a small bistro table against the wall. She is a middle-aged woman wearing a rabbit-skin bolero vest and a full skirt. What I'm really wondering is why notices were posted and invitations sent out at all: If it's a psychics' convention, shouldn't everyone just *know* where and when to meet?

Taking my hand, she rakes it lightly with her spread fingers, then lifts it up close to her face as if zeroing in on a splinter.

"You drive a red car," she says in a solemn voice.

"No, a blue one . . . ," I say, hating to disappoint her.

"Well you *will* drive a red car in the future sometime, and you must be very careful," she warns. "I see a lot of money for you in December, but someone you work with will betray you, and you must watch out. . . . You're close to someone named Mary?"

I shake my head no.

"Margaret? Melissa? Monica?"

"I have a mother named Marcia," I offer.

"Ah, that's it, and you're very concerned about her, but she'll be all right, you don't have to worry." Now she presses the fleshy side of my palm and folds back the thumb, separates the fingers and peers closely at them. The hand is "the visible part of the brain," Immanuel Kant once said. She searches the *flexure lines* (creases made by moving the hand), *tension lines* (wrinkles that grow with age the way facial lines do), and *papillary ridges* (fingerprints), traces my head line, heart line, life line and fate line. Among our near neighbors, the apes, the heart and head lines are the same, but so mobile and powerful are our forefingers that they tend to separate the lines on most people. My hands are cool and dry. Palms sweat when we're agitated, in tribute to a time deep in our past when stress meant physical danger and our body wanted us ready to fight or flee. A tiny discoloration at the base of my second finger brings a nod of interest from the palm-reader. It's only a scar left from a rose thorn, nothing like stigmata, marks some Roman Catholics claim appear spontaneously on their feet and palms and bleed, reproducing the wounds Christ suffered on the cross.

"You know someone who had an abortion?" the palm-reader asks.

Throughout history, palm-readers have chosen the hand as their symbolic link to the psyche and soul, as their raft through time. After all, the hand is action, it digs roads and builds cities, it throws spears and diapers babies. Even its small dramas—dialing a phone number, pushing a button—can change the course of nations or launch atomic bombs. When we are distressed, we allow our hands to console each other by wringing, stroking, fidgeting, and caressing them as if they were separate people. At the outset of a romance, the first touch people share is usually the taking of each other's hand, while couples of long standing, moving through the world on their daily rounds, often hold hands as a tender bridge. Holding the hand of someone ill or elderly soothes them and gives them an emotional lifeline. Experiments show that just touching someone's hand or arm lowers their blood pressure. In many cultures, people fiddle obsessively with worry beads, polished stones, and other objects, and the brain-wave patterns this produces are those of a mind made calm by repeated touch stimulation.

In these days of mass-produced objects, we treasure things that are "handmade." We think of manual laborers as working harder than desk jockeys, though it might not always be the case. Sometimes working hands seem to perform with a cunning and sensitivity that defies explanation. Lorraine Miller, though totally blind, works as a hair stylist at a beauty salon in Lancaster, Pennsylvania. A mother of five, Ms. Miller had always wanted to be a beautician, but the rigors of raising a family never allowed time for it. Later in life, blinded by disease, she decided to pursue her lifelong ambition. A hair salon in Lancaster, Pennsylvania, trained her to cut hair by touch, carefully feeling the shape of the head and the layers of hair as she cut them. In time, she touch-cut so well that they hired her.

The tiny ridges in our fingertips, whose roughness makes it easier for us to grasp objects, are randomly formed, resulting in the unique swirling weather systems we call "fingerprints." The swirls run through a few basic patterns of whorls, loops, and arches, but combine in endlessly different ways. Not even identical twins have the same fingerprints, which makes guilt a lot easier to establish when it is necessary to do so. The idea of one's fingerprints being the ultimate personal signature isn't new. Thousands of years ago, the Chinese used the imprint of a finger as a way of signing a contract. When the FBI searches for fingerprints on a holdup note, they use a laser. The oily residue absorbs laser light and re-emits it at a longer wavelength. Forensic experts wearing amber goggles then filter out the laser light and see the fingerprints—always a distinctive signature.

A hand moves with a complex precision that's irreplaceable, feels with a delicate intuition that's indefinable, as designers of robotic hands are discovering. Because we use our hands so often for so many purposes, flexing, bending, gripping, pointing, stretching them millions of times, University of Utah Research Institute engineers have invented a glove to wear over a hand that has lost the sense of touch—through the use of electronics and sound waves, it gives the wearer a sense of pressure, which is essential to being able to grasp. A wire leads from the glove to a tiny piston that is connected to a part of the body where feeling hasn't been lost, and the wearer feels hand sensations (in his wrist or forearm, for example) and learns to translate them into hand responses.

The sensitivity of the fingertips reveals itself in the use of Braille, which now appears everywhere, from elevator panels to the faces of Italian coins. Braille can be read quickly, and people are always looking for better ways to use it. A recent study reported in *Education of the Visually Handicapped* suggests that Braille can be read more accurately and efficiently if the readers move their fingers vertically over the dots rather than horizontally, because the fingertip's touch receptors are more sensitive when used in that way.

Handclasps and handshakes have served throughout history to prove the lack of a weapon and to pledge one's good faith, although shaking hands as a common greeting didn't really come into practice until the Industrial Revolution in England, when businessmen were so busy making deals and shaking hands on them that the gesture lost its special purpose and entered casual social life. A handshake is still a watered-down contract that says: Let's at least pretend that we'll deal honorably with each other. The hand has been symbolic of the whole body for some time, as in "I'll give you a hand," or referring to a worker as a "hired hand."

Think of all the ways in which we touch ourselves (I don't just mean masturbation—from *manustuprare*, "to defile with the hand"), but how we wrap our hands around our shoulders and rock as if we were a mother comforting a child; how we hide our face in our open palms to be alone to pray, or that they may receive our tears; how we run our hands briskly up and down our arms as we pace; how, with wide eyes, we press an open palm to one cheek when we're startled. Touch is so important in emotional situations that we're driven to touch ourselves in the way we'd like someone else to comfort us. Hands are messengers of emotion. And few have understood their intricate duty as well as Rodin. Here is how Rilke describes Rodin's artistry:

Rodin has made hands, independent, small hands which, without forming part of the body, are yet alive. Hands rising upright, angry and irritated, hands whose five bristling fingers seem to bark like the five throats of Cerberus. Hands in motion, sleeping hands and hands in the act of awakening; criminal hands weighted by heredity, hands that are tired and have lost all desire, lying like some sick beast crouched in a corner, knowing none can help them. But hands are a complicated organism, a delta in which much life from distant sources flows together and is poured into the great stream of action. Hands have a history of their own, they have indeed, their own civilization, their special beauty; we concede to them the right to have their own development, their own wishes, feelings, moods and favorite occupations.

PROFESSIONAL TOUCHERS

In the sea of so-called healers who cater to desperate people, there are practitioners of "therapeutic touch," who claim to cure people of physical ills without actually touching the body, by running their hands at a discreet distance over a person's energy field. The ancient practice of "the laying on of hands" can be seen weekly on most TV sets in the United States. A preacher calls a sick or troubled person out of the audience, seems to intuit their problem without being told (charlatan-debunker Randi has revealed simple magician's tricks that are used), and then touches them on the forehead with such force it knocks them off their feet. They fall to the ground in religious ecstasy, stand up and claim to be healed. Throughout the world, shamans and medicine men perform similar rituals, seeming to draw the demon out of a person's body, healing them with an incantation and a touch.

Touch is so powerful a healer that we go to professional touchers (doctors, hairdressers, masseuses, dancing instructors, cosmeticians, barbers, gynecologists, chiropodists, tailors, back manipulators, prostitutes, and manicurists), and frequent emporiums of touch—discothèques, shoeshine stands, mud baths. Illness usually sends us to a doctor, but often we go just to be fussed over and touched. A doctor can't help much when one has a minor allergy, the flu, or some other small affliction, but we go anyway to be patted, stroked, listened to, inspected, handled. Monkeys and other animals engage in a lot of grooming, especially of the head. The ancient Romans, Greeks, and Egyptians wore elaborate coifs that required the steady attendance of hairdressers, but this voluptuous touching eventually went out of fashion and didn't reappear until after the Middle Ages; the professional beauty salon didn't come into vogue until the Victorian era.

Gynecologists do the most intimate professional touching of all, and few situations are as awkward for a woman as having a male gynecologist she's never so much as said hello to walk into an examining room, lift up the sheet, and set to work. Such a blasé attitude hasn't always been the hallmark of a gynecologist's calling. "Three hundred years ago he was even on occasion required to crawl into the pregnant woman's bedroom on his hands and knees to perform the examination," Desmond Morris observes, "so that she would be unable to see the owner of the fingers which were to touch her so privately. At a later date, he was forced to work in a darkened room, or to deliver a baby by groping beneath the bedclothes. A 17th-century etching shows him sitting at the foot of the labour bed with the sheet tucked into his collar like a napkin, so that he is unable to see what his hands are doing, an anti-intimacy device that made cutting the umbilical cord a particularly hazardous operation."

The most obvious professional touch is the massage, designed to stimulate circulation, dilate blood vessels, relax tense muscles, and clean toxins out of the body through the flow of lymph. The popular "Swedish" massage emphasizes long, sweeping strokes in the direction of the heart. The Japanese "shiatsu" is a kind of acapuncture without needles, using the finger (*shi* in Japanese) to cause pressure *(atsu)*. The body is charted according to meridians, along which one's vitality or life-force flows, and the massage frees the way for it. In "neo-Reichian" massage, which is sometimes used in conjunction with psychotherapy, the practitioner strokes away from the heart in order to dispel nervous energy. "Reflexology" focuses on the feet, but, like shiatsu, also attends to pressure points on the skin, which represent various organs. Massaging these points is supposed to help the corresponding organ to function better. In "Rolfing," the massage turns into violent, sometimes painful manipulation. Although there are many different massage techniques, some formal schools, and much philosophizing on the subject, studies have shown that loving touching alone—in whatever style—can improve health.

At Ohio University School, one researcher conducted an experiment in which he fed rabbits high-cholesterol diets and methodically petted a special group of them; the petted rabbits had a 50 percent lower rate of arteriosclerosis than similarly fed but unpetted rabbits.

A Philadelphia experiment studied the survival chances of patients who had had heart attacks. Examining a wide spectrum of variables and their effects on survival, the experiments discovered that the variable that produced the strongest effect was pet ownership. It made no difference if the person were married or single—pet owners still survived the longest. The idle stroking of our pets that is so calming and can be done almost subconsciously while we do something else or talk to friends or work has a healing effect. As one of the experimenters said: "We raise our children in a nontactile society and have to compensate with nonhuman creatures. First with teddy bears and blankets, then with pets. When touch isn't there, our true isolation comes through." Touching is just as therapeutic as being touched; the healer, the giver of touch, is simultaneously healed.

TABOOS

Despite our passion, indeed our need, to touch and be touched, many parts of the body are taboo in different cultures. In the United States, it isn't acceptable for a man to touch the breasts, buttocks, or genitals of a woman who doesn't invite him to do so. Because a woman tends to be shorter than a man, when he puts an arm around her shoulder her arm falls naturally around his waist. As a result, a woman often ends up touching a man's waist and pelvis without its becoming a necessarily sexual act. When a man touches a woman's pelvis, though, it immediately registers as sexual. Women touch other women's hair and faces more often than men touch other men's hair and faces. Females, in general, have their hair touched more by everyone—mothers, fathers, boyfriends, girlfriends—than males do. It's taboo to touch a Japanese girl's nape. In Thailand, it's taboo to touch the top of a girl's head. In Fiji, touching someone's hair is as taboo as touching the genitals of a stranger would be in, say, Iowa. Even primitive tribes, in which men and women walk around naked, have taboos about touching parts of the body. In fact, there are only two situations when the taboos disappear: Lovers have complete access to the body of another person, and so does a mother with her baby. Many of the encounter groups that blossomed during the sixties were little more than organized touch sessions, often "aided" by drugs, in which people tried to break down some of the social inhibitions and taboos that left them feeling pent-up, rigid, and alien.

There are also gender and status taboos. We look at, talk with, and listen to all sorts of people every day of our lives, but touch is special. Touching someone is like using their first name. Think about two people talking in a business meeting: One of them touches the other lightly on the hand while making a point, or puts an arm around the other's shoulder. Which one is the boss? The one who initiates a touch is almost always the person of higher status. Researchers observing hundreds of people in public settings in a small town in Indiana and in a big city on the East Coast, found that males touch females first, that females are more likely to touch females than males are to touch males, and that people of higher status generally touch lower-status people first. Lower-status people wait for the go-ahead before they risk an increased intimacy—even a subconscious one—with their presumed superiors.

SUBLIMINAL TOUCH

At Purdue University Library, a woman librarian goes about her business, checking out people's books. She is part of an experiment in subliminal touch, and knows that half the time she is to do nothing special, the other half to touch people as insignificantly as possible. She brushes a student's hand lightly as she returns a library card. Then the student is followed outside and asked to fill out a questionnaire about the library that day. Among other questions, the student is asked if the librarian smiled, and if she touched him. In fact, the librarian had not smiled, but the student reports that she did, although he says she did not touch him. This experiment lasts all day, and soon a pattern becomes clear: those students who have been subconsciously touched report much more satisfaction with the library and life in general.

In a related experiment staged at two restaurants in Oxford, Mississippi, waitresses lightly and unobtrusively touch diners on the hand or shoulder. Those customers who are touched don't necessarily rate the food or restaurant better, but they consistently tip the waitress higher. In yet another experiment in Boston, a researcher leaves money in a phone booth, then returns when she sees the next person pocket the money; she casually asks if they've found what she lost. If the researcher touches the person while asking for their help, touches them insignificantly so that they don't remember it later, the likelihood that the money will be returned rises from 63 to 96 percent. Despite the fact that we're territorial creatures who move through the world like small principalities, contact warms us even without our knowing it. It probably reminds us of that time, long before deadlines and banks, when our mothers cradled us and we were enthralled and felt perfectly lovable. Even touch so subtle as to be overlooked doesn't go unnoticed by the subterranean mind.

Taste

*Those... from whom nature has
withheld the legacy of taste,
have long faces, and long
eyes and noses, whatever their
height there is something elongated
in their proportions. Their hair
is dark and unglossy, and they are
never plump; it was they
who invented trousers.*

Anthelme Brillat-Savarin,
The Physiology of Taste

THE SOCIAL SENSE

The other senses may be enjoyed in all their beauty when one is alone, but taste is largely social. Humans rarely choose to dine in solitude, and food has a powerful social component. The Bantu feel that exchanging food makes a contract between two people who then have a "clanship of porridge." We usually eat with our families, so it's easy to see how "breaking bread" together would symbolically link an outsider to a family group. Throughout the world, the stratagems of business take place over meals; weddings end with a feast; friends reunite at celebratory dinners; children herald their birthdays with ice cream and cake; religious ceremonies offer food in fear, homage, and sacrifice; wayfarers are welcomed with a meal. As Brillat-Savarin says, "every . . . sociability . . . can be found assembled around the same table: love, friendship, business, speculation, power, importunity, patronage, ambition, intrigue . . ." If an event is meant to matter emotionally, symbolically, or mystically, food will be close at hand to sanctify and bind it. Every culture uses food as a sign of approval or commemoration, and some foods are even credited with supernatural powers, others eaten symbolically, still others eaten ritualistically, with ill fortune befalling dullards or skeptics who forget the recipe or get the order of events wrong. Jews attending a Seder eat a horseradish dish to symbolize the tears shed by their ancestors when they were slaves in Egypt. Malays celebrate important events with rice, the inspirational center of their lives. Catholics and Anglicans take a communion of wine and wafer. The ancient Egyptians thought onions symbolized the many-layered universe, and swore oaths on an onion as we might on a Bible. Most cultures embellish eating with fancy plates and glasses, accompany it with parties, music, dinner theater, open-air barbecues, or other forms of revelry. Taste is an intimate sense. We can't taste things at a distance. And how we taste things, as well as the exact makeup of our saliva, may be as individual as our fingerprints.

Food gods have ruled the hearts and lives of many peoples. Hopi Indians, who revere corn, eat blue corn for strength, but all Americans might be worshiping corn if they knew how much of their daily lives depended on it. Margaret Visser, in *Much Depends on Dinner*, gives us a fine history of corn and its uses: livestock and poultry eat corn; the liquid in canned foods contains corn; corn is used in most paper products, plastics, and adhesives; candy, ice cream, and other goodies contain corn syrup; dehydrated and instant foods contain cornstarch; many familiar objects are made from corn products, brooms and corncob pipes to name only two. For the Hopis, eating corn is itself a form of reverence. I'm holding in my hand a beautifully carved Hopi corn kachina doll made from cottonwood; it represents one of the many spiritual essences of their world. Its cob-shaped body is painted ocher, yellow, black, and white, with dozens of squares drawn in a cross-section-of-a-kernel design, and abstract green leaves spearing up from below. The face has a long, black, rootlike nose, rectangular black eyes, a black ruff made of rabbit fur, white string corn-silk-like ears, brown bird-feather bangs, and two green, yellow, and ocher striped horns topped by rawhide tassels. A fine, soulful kachina, the ancient god Maïs stares back at me, tastefully imagined.

Throughout history, and in many cultures, *taste* has always had a double meaning. The word comes from the Middle English *tasten*, to examine by touch, test, or sample, and continues back to the Latin *taxare*, to touch sharply. So a taste was always a trial or test. People who have taste are those who have appraised life in an intensely personal way and found some of it sublime, the rest of it lacking. Something in bad taste tends to be obscene or vulgar. And we defer to professional critics of wine, food, art, and so forth, whom we trust to taste things for us because we think their taste more refined or educated than ours. A companion is "one who eats bread with another," and people sharing food as a gesture of peace or hospitality like to sit around and chew the fat.

The first thing we taste is milk from our mother's breast,* accompanied by love and affection, stroking, a sense of security, warmth, and well-being, our first intense feelings of pleasure. Later on she will feed us solid food from her hands, or even chew food first and press it into our mouths, partially digested. Such powerful associations do not fade easily, if at all. We say "food" as if it were a simple thing, an absolute like rock or rain to take for granted. But it is a big source of pleasure in most lives, a complex realm of satisfaction both physiological and emotional, much of which involves memories of childhood. Food must taste good, must reward us, or we would not stoke the furnace in each of our cells. We must eat to live, as we must breathe. But breathing is involuntary, finding food is not; it takes energy and planning, so it must tantalize us out of our natural torpor. It must decoy us out of bed in the morning and prompt us to put on constricting clothes, go to work, and perform tasks we may not enjoy for eight hours a day, five days a week, just to "earn our daily bread," or be "worth our salt," if you like, where the word *salary* comes from. And, because we are omnivores, many tastes must appeal to us, so that we'll try new foods. As children grow, they meet regularly throughout the day—at mealtimes—to hear grown-up talk, ask questions, learn about customs, language, and the world. If language didn't arise at mealtimes, it certainly evolved and became more fluent there, as it did during group hunts.

We tend to see our distant past through a reverse telescope that compresses it: a short time as hunter-gatherers, a long time as "civilized" people. But civilization is a recent stage of human life, and, for all we know, it may not be any great achievement. It may not even be the final stage. We have been alive on this planet as recognizable humans for about two million years, and for all but the last two or three thousand we've been hunter-gatherers. We may sing in choirs and park our rages behind a desk, but we patrol the world

*This special milk, called colostrum, is rich in antibodies, the record of the mother's epidemiologic experience.

with many of a hunter-gatherer's drives, motives, and skills. These aren't knowable truths. Should an alien civilization ever contact us, the greatest gift they could give us would be a set of home movies: films of our species at each stage in our evolution. Consciousness, the great poem of matter, seems so unlikely, so impossible, and yet here we are with our loneliness and our giant dreams. Speaking into the perforations of a telephone receiver as if through the screen of a confessional, we do sometimes share our emotions with a friend, but usually this is too disembodied, too much like yelling into the wind. We prefer to talk *in person,* as if we could temporarily slide into their feelings. Our friend first offers us food, drink. It is a symbolic act, a gesture that says: *This food will nourish your body as I will nourish your soul.* In hard times, or in the wild, it also says *I will endanger my own life by parting with some of what I must consume to survive.* Those desperate times may be ancient history, but the part of us forged in such trials accepts the token drink and piece of cheese and is grateful.

FOOD AND SEX

What would the flutterings of courtship be without a meal? As the deliciously sensuous and ribald tavern scene in Fielding's *Tom Jones* reminds us, a meal can be the perfect arena for foreplay. Why is food so sexy? Why does a woman refer to a handsome man as a real dish? Or a French girl call her lover *mon petit chou* (my little cabbage)? Or an American man call his girlfriend cookie? Or a British man describe a sexy woman as a bit of crumpet (a flat, toasted griddlecake well lubricated with butter)? Or a tart? Sexual hunger and physical hunger have always been allies. Rapacious needs, they have coaxed and driven us through famine and war, to bloodshed and serenity, since our earliest days.

Looked at in the right light, any food might be thought aphrodisiac. Phallic-shaped foods such as carrots, leeks, cucumbers, pickles, sea cucumbers (which become tumescent when soaked), eels, bananas, and asparagus all have been prized as aphrodisiacs at one time or another, as were oysters and figs because they reminded people of female genitalia; caviar because it was a female's eggs; rhinoceros horn, hyena eyes, hippopotamus snout, alligator tail, camel hump, swan genitals, dove brains, and goose tongues, on the principle that anything so rare and exotic must have magical powers; prunes (which were offered free in Elizabethan brothels); peaches (because of their callipygous rumps?); tomatoes, called "love apples," and thought to be Eve's temptation in the Garden of Eden; onions and potatoes, which look testicular, as well as "prairie oysters," the cooked testicles of a bull; and mandrake root, which looks like a man's thighs and penis. Spanish fly, the preferred aphrodisiac of the Marquis de Sade, with which he laced the bonbons he fed prostitutes and friends, is made by crushing a southern European beetle. It contains a gastrointestinal irritant and also produces a better blood flow, the combination of which brings on a powerful erection of either the penis or the clitoris, but also damages the kidneys; it can even be fatal. Musk, chocolate, and truffles also have been considered aphrodisiac and, for all we know, they might well be. But, as sages have long said, the sexiest part of the body and the best aphrodisiac in the world is the imagination.

Primitive peoples saw creation as a process both personal and universal, the earth's yielding food, humans (often molded from clay or dust) burgeoning with children. Rain falls from the sky and impregnates the ground, which brings forth fruit and grain from the tawny flesh of the earth—an earth whose mountains look like reclining women, and whose springs spurt like healthy men. Fertility rituals, if elaborate and frenzied enough, could encourage Nature's bounty. Cooks baked meats and breads in the shape of genitals, especially penises, and male and female statues with their sexual organs exaggerated presided over orgiastic festivities where sacred couples copulated in public. A mythic Gaia poured milk from her breasts and they became the galaxies. The ancient Venus figures with global breasts, swollen bellies, and huge buttocks and thighs symbolized the female life-force, mother to crops and humans. The earth itself was a goddess, curvy and ripe, radiant with fertility, aspill with riches. People have thought the Venus figures imaginative exaggerations, but women of that time may indeed have resembled them, all breasts, belly, and rump. When pregnant, they would have bulged into quite an array of shapes.

Food is created by the sex of plants or of animals; and we find it sexy. When we eat an apple or peach, we are eating the fruit's placenta. But, even if that weren't so, and we didn't subconsciously associate food with sex, we would still find it sexy for strictly physical reasons. We use the mouth for many things—to talk and kiss, as well as to eat. The lips, tongue, and genitals all have the same neural receptors, called Krause's end bulbs, which make them ultrasensitive, highly charged. There's a similarity of response.

A man and woman sit across from one another in a dimly lit restaurant. A small bouquet of red-and-white spider lilies sweetens the air with a cinnamonlike tingle. A waiter passes with a plate of rabbit sausage in molé sauce. At the next table, a blueberry soufflé oozes scent. Oysters on the half shell, arranged on a large platter of shaved ice, one by one polish the woman's tongue with silken saltiness. A fennel-scented steam rises from thick crabcakes on the man's plate. Small loaves of fresh bread breathe sweetly. Their hands brush as they both reach for the bread. He stares into her eyes, as if filling them with molten lead. They both know where this delicious prelude will lead. *"I'm so hungry,"* she whispers.

THE OMNIVORE'S PICNIC

You have been invited to dinner at the home of extraterrestrials, and asked to bring friends. Being considerate hosts, they first inquire if you have any dietary allergies or prohibitions, and then what sort of food would taste good to you. What do humans eat? they ask. Images cascade through your mind, a cornucopia of plants, animals, minerals, liquids, and solids, in a vast array of cuisines. The Masai enjoy drinking cow's blood. Orientals eat stir-fried puppy. Germans eat rancid cabbage (sauerkraut), Americans eat decaying cucumbers (pickles), Italians eat whole deep-fried songbirds, Vietnamese eat fermented fish dosed with chili peppers, Japanese and others eat fungus (mushrooms), French eat garlic-soaked snails. Upper-class Aztecs ate roasted dog (a hairless variety named *xquintli*, which is still bred in Mexico). Chinese of the Chou dynasty liked rats, which they called "household deer,"* and many people still do eat rodents, as well as grasshoppers, snakes, flightless birds, kangaroos, lobsters, snails, and bats. Unlike most other animals, which fill a small yet ample niche in the large web of life on earth, humans are omnivorous. The Earth offers perhaps 20,000 edible plants alone. A poor season for eucalyptus will wipe out a population of koala bears, which have no other food source. But human beings are Nature's great ad libbers and revisers. Diversity is our delight. In time of drought, we can ankle off to a new locale, or break open a cactus, or dig a well. When plagues of locusts destroy our crops, we can forage on wild plants and roots. If our herds die, we find protein in insects, beans, and nuts. Not that being an omnivore is easy. A koala bear doesn't have to worry about whether or not its next mouthful will be toxic. In fact, eucalyptus is highly poisonous, but a koala has an elaborately protective gut, so it just eats eucalyptus, exactly as its parents did. Cows graze without fear on grass and grain. But omnivores are anxious eaters. They must continually test new foods to see if they're palatable and nutritious, running the risk of inadvertently poisoning themselves. They must take chances on new flavors, and, doing so, they frequently acquire a taste for something offbeat that, though nutritious, isn't the sort of thing that might normally appeal to them—chili peppers (which Columbus introduced to Europe), tobacco, alcohol, coffee, artichokes, or mustard, for instance. When we were hunter-gatherers, we ate a great variety of foods. Some of us still do, but more often we add spices to what we know, or find at hand, *for variety*, as we like to say. Monotony isn't our code. It's safe, in some ways, but in others it's more dangerous. Most of us prefer our foods cooked to the steaminess of freshly killed prey. We don't have ultrasharp carnivore's teeth, but we don't need them. We've created sharp tools. We do have incisor teeth for slicing fruits, and molars for crushing seeds and nuts, as well as canines for ripping

---

*It was the food-obsessed Chinese who started the first serious restaurants during the time of the T'ang dynasty (A.D. 618–907). By the time the Sung dynasty replaced the T'ang, they were all-purpose buildings, with many private dining rooms, where one went for food, sex, and barroom gab.

flesh. At times, we eat nasturtiums and pea pods and even the effluvia from the mammary glands of cows, churned until it curdles, or frozen into a solid and attached to pieces of wood.

Our hosts propose a picnic, since their backyard is a meadow lit by two suns, and they welcome us and our friends. Our Japanese friend chooses the appetizer: sushi, including shrimp still alive and wriggling. Our French friend suggests a baguette, or better still croissants, which have an unlikely history, which he insists on telling everyone: To celebrate Austria's victory against the invading Ottoman Turks, bakers created pastry in the shape of the crescent on the Turkish flag, so that the Viennese could devour their enemies at table as they had on the battlefield. Croissants soon spread to France and, during the 1920s, traveled with other French ways to the United States. Our Amazonian friend chooses the main course— nuptial kings and queens of leaf-cutter ants, which taste like walnut butter, followed by roasted turtle and sweet-fleshed piranha. Our German friend insists that we include some spaetzle and a loaf of darkest pumpernickel bread, which gets its name from the verb *pumpern*, "to break wind," and *Nickel*, "the devil," because it was thought to be so hard to digest that even the devil would fart if he ate it. Our Tasaday friend wants some natek, a starchy paste his people make from the insides of caryota palm trees. The English cousin asks for a small platter of potted ox tongues, very aged blue cheese, and, for dessert, trifle—whipped cream and slivered almonds on top of a jam-and-custard pudding thick with sherry-soaked lady-fingers.

To finish our picnic lunch, our Turkish friend proposes coffee in the Turkish style—using a mortar and pestle to break up the beans, rather than milling them. To be helpful, he prepares it for us all, pouring boiling water over coffee grounds through a silver sieve into a pot. He brings this to a light boil, pours it through the sieve again, and offers us some of the clearest, brightest coffee we've ever tasted. According to legend, he explains, coffee was discovered by a ninth-century shepherd, who one day realized that his goats were becoming agitated whenever they browsed on the berries of certain bushes. For four hundred years, people thought only to chew the berries.

Raw coffee doesn't brew into anything special, but in the thirteenth century someone decided to roast the berries, which releases a pungent oil and the mossy-bitter aroma now so familiar to us. Our Indian friend passes round cubes of sugar, which we are instructed to let melt on the tongue as we sip our coffee, and our minds roam back to the first recorded instance of sugar, in the Atharvaveda, a sacred Hindu text from 800 B.C., which describes a royal crown made of glittering sugar crystals. Then he circulates a small dish of coriander seeds, and we pinch a few in our fingers, set them on our tongues, and feel our mouths freshen from the aromatic tang. A perfect picnic. We thank our hosts for laying on such a splendid feast, and invite them to our house for dinner next. "What do jujubarians eat?" we ask.

OF CANNIBALISM AND SACRED COWS

Even though grass soup was the main food in the Russian gulags, according to Solzhenitsyn's *One Day in the Life of Ivan Denisovich*, humans don't prefer wood, or leaves, or grass—the cellulose is impossible to digest. We also can't manage well eating excrement, although some animals adore it, or chalk or petroleum. On the other hand, cultural taboos make us spurn many foods that are wholesome and nourishing. Jews don't eat pork, Hindus don't eat beef, and Americans in general won't eat dog, rat, horse, grasshopper, grubs, or many other palatable foods prized by peoples elsewhere in the world. Anthropologist Claude Lévi-Strauss found that primitive tribes designated foods "good to think" or "bad to think." Necessity, the mother of invention, fathers many codes of conduct. Consider the "sacred cow," an idea so shocking it has passed into our vocabulary as a thing, event, or person considered sacrosanct. Though India has a population of around 700 million and a constant need for protein, over two hundred million cattle are allowed to roam the streets as deities while many people go hungry. The cow plays a central role in Hinduism. As Marvin Harris explains in *The Sacred Cow and the Abominable Pig*:

Cow protection and cow worship also symbolize the protection and adoration of human motherhood. I have a collection of colorful Indian pin-up calendars depicting jewel-bedecked cows with swollen udders and the faces of beautiful human madonnas. Hindu cow worshippers say: "The cow is our mother. She gives us milk and butter. Her male calves till the land and give us food." To critics who oppose the custom of feeding cows that are too old to have calves and give milk, Hindus reply: "Will you then send your mother to a slaughter house when she gets old?"

Not only is the cow sacred in India, even the dust in its hoofprints is sacred. And, according to Hindu theology, 330 million gods live inside each cow. There are many reasons why this national tantalism has come about; one factor may be that an overcrowded land such as India can't support the raising of livestock for food, a system that is extremely inefficient. When people eat animals that have been fed grains, "nine out of ten calories and four out of five grams of protein are lost for human consumption." The animal uses up most of the nutrients. So vegetarianism may have evolved as a remedy, and been ritualized through religion. "I feel confident that the rise of Buddhism was related to mass suffering and environmental depletions," Harris writes, "because several similar nonkilling religions . . . arose in India at the same time." Including Jainism, whose priests not only tend stray cats and dogs, but keep a separate room in their shelters just for insects. When they walk down the street, an assistant walks ahead of them to brush away any insects lest they get stepped on, and they wear gauze masks so they don't accidentally inhale a wayward midge or other insect.

One taboo stands out as the most fantastic and forbidden. "What's eating you?" a man may ask an annoyed friend. Even though his friend just got fired by a tyrannical boss with a mind as small as a noose, he would never think to say "*Who's* eating you?" The idea of cannibalism is so far from our ordinary lives that we can safely use the euphemism *eat* in a sexual context, say, and no one will think we mean literally consume. But omnivores can eat anything, even each other,* and human flesh is one of the finest sources of protein. Primitive peoples all over the world have indulged in cannibalism, always ritualistically, but sometimes as a key source of protein missing from their diets. For many it's a question of head-hunting, displaying the enemy's head with much magic and flourish; and then, so as not to be wasteful, eating the body. In Britain's Iron Age, the Celts consumed large quantities of human flesh. Some American Indian tribes tortured and ate their captives, and the details (reported by Christian missionaries who observed the rites) are hair-raising. During one four-night celebration in 1487, the Aztecs were reported to have sacrificed about eighty thousand prisoners, whose flesh was shared with the gods, but mainly eaten by a huge meat-hungry population. In *The Power of Myth,* the late Joseph Campbell, a wise observer of the beliefs and customs of many cultures, tells of a New Guinea cannibalism ritual that "enacts the planting-society myth of death, resurrection and *cannibalistic* consumption." The tribe enters a sacred field, where they chant and beat drums for four or five days, and break all the rules by engaging in a sexual orgy. In this rite of manhood, young boys are introduced to sex for the first time:

> There is a great shed of enormous logs supported by two uprights. A young woman comes in ornamented as a deity, and she is brought to lie down in this place beneath the great roof. The boys, six or so, with the drums going and chanting going, one after another, have their first experience of intercourse with the girl. And when the last boy is with her in full embrace, the supports are withdrawn, the logs drop, and the couple is killed. There is the union of male and female . . . as they were in the beginning. . . . There is the union of begetting and death. They are both the same thing.
>
> Then the couple is pulled out and roasted and eaten that very evening. The ritual is the repetition of the original act of the killing of a god followed by the coming of food from the dead savior.

---

*In German, humans eat *(essen),* but animals devour or feed *(fressen).* Cannibals are called *Menschenfresser*—humans who become animals when they eat.

When the explorer Dr. Livingstone died in Africa, his organs were apparently eaten by two of his native followers as a way to absorb his strength and courage. Taking communion in the Catholic Church enacts a symbolic eating of the body and blood of Christ. Some forms of cannibalism were more bloodthirsty than others. According to Philippa Pullar, Druid priests "attempted divination by stabbing a man above his midriff, foretelling the future by the convulsions of his limbs and the pouring of his blood. . . . Then . . . they devoured him." Cannibalism doesn't horrify us because we find human life sacred, but because our social taboos happen to forbid it, or, as Harris says: "the real conundrum is why we who live in a society which is constantly perfecting the art of mass-producing human bodies on the battlefield find humans good to kill but bad to eat."*

THE BLOOM OF A TASTE BUD

Seen by scanning electron microscope, our taste buds look as huge as volcanoes on Mars, while those of a shark are beautiful mounds of pastel-colored tissue paper—until we remember what they're used for. In reality, taste buds are exceedingly small. Adults have about 10,000, grouped by theme (salt, sour, sweet, bitter), at various sites in the mouth. Inside each one, about fifty taste cells busily relay information to a neuron, which will alert the brain. Not much tasting happens in the center of the tongue, but there are also incidental taste buds on the palate, pharynx, and tonsils, which cling like bats to the damp, slimy limestone walls of a cave. Rabbits have 17,000 taste buds, parrots only about 400, cows 25,000. What are they tasting? Maybe a cow needs that many to enjoy a relentless diet of grass.

At the tip of the tongue, we taste sweet things; bitter things at

---

*For an excellent discussion of cannibalism, and the nutritional fiats that have prompted it in a variety of cultures (Aztecs, Fijians, New Guineans, American Indians, and many others), including truly horrible and graphic accounts by eyewitnesses, see Harris's chapter on "People Eating."

the back; sour things at the sides; and salty things spread over the surface, but mainly up front. The tongue is like a kingdom divided into principalities according to sensory talent. It would be as if all those who could see lived to the east, those who could hear lived to the west, those who could taste lived to the south, and those who could touch lived to the north. A flavor traveling through this kingdom is not recognized in the same way in any two places. If we lick an ice cream cone, a lollipop, or a cake-batter-covered finger, we touch the food with the tip of the tongue, where the taste buds for sweetness are, and it gives us an extra jolt of pleasure. A cube of sugar under the tongue won't taste as sweet as one placed *on* the tongue. Our threshold for bitter is the lowest. Because the taste buds for bitter lie at the back of the tongue; as a final defense against danger they can make us gag to keep a substance from sliding down the throat. Some people do, in fact, gag when they take quinine, or drink coffee for the first time, or try olives. Our taste buds can detect sweetness in something even if only one part in two hundred is sweet. Butterflies and blowflies, which have most of their taste organs on their front feet, need only step in a sweet solution to taste it. Dogs, horses, and many other animals have a sweet tooth, as we do. We can detect saltiness in one part in 400, sourness in one part in 130,000, but bitterness in as little as one part in 2,000,000. Nor is it necessary for us to recognize poisonous things as tasting different from one another; they just taste bitter. Distinguishing between bitter and sweet substances is so essential to our lives that it has burst through our language. Children, joy, a trusted friend, a lover all are referred to as "sweet." Regret, an enemy, pain, disappointment, a nasty argument all are referred to as "bitter." The "bitter pill" we metaphorically dread is likely to be poison.

Taste buds got their name from the nineteenth-century German scientists Georg Meissner and Rudolf Wagner, who discovered mounds made up of taste cells that overlap like petals. Taste buds wear out every week to ten days, and we replace them, although not as frequently over the age of forty-five—our palates really do become jaded* as we get older. It takes a more intense taste to produce the same level of sensation, and children have the keenest sense of taste. A baby's mouth has many more taste buds than an adult's, with some even dotting the cheeks. Children adore sweets partly because the tips of their tongues, more sensitive to sugar, haven't yet been blunted by years of gourmandizing or trying to eat hot soup before it cools. A person born without a tongue, or who has had his tongue cut out, still can taste. Brillat-Savarin tells of a Frenchman in Algeria who was punished for an attempted prison escape by having "the forepart of his tongue . . . cut off clear to the ligament." Swallowing was difficult and tiring for him, although he could still taste fairly well, "but very sour or bitter things caused him unbearable pain."

Just as we can smell something only when it begins to evaporate, we can taste something only when it begins to dissolve, and we cannot do that without saliva. Every taste we can imagine—from mangoes to hundred-year-old eggs—comes from a combination of the four primary tastes plus one or two others. And yet we can distinguish between tastes with finesse, as wine-, tea-, cheese- and other professional tasters do. The Greeks and Romans, who were sophisticated about fish, could tell just by tasting one what waters it came from. As precise as our sense of taste is, illusions can still surprise us. For example, MSG doesn't taste saltier than table salt, but it really contains much more sodium. One of its ingredients, glutamate, blocks our ability to taste it as salty. A neurologist at the Albert Einstein College of Medicine once tested the amount of MSG in a bowl of wonton soup in a Chinese restaurant in Manhattan, and he found 7.5 grams of MSG, as much sodium as one should limit oneself to in an entire day.

After brushing our teeth in the morning, orange juice tastes bitter. Why? Because our taste buds have membranes that contain fatlike phospholipids, and toothpastes contains a detergent that breaks down fat and grease. So the toothpaste first assaults the membranes with its detergent, leaving them raw; then chemicals in the toothpaste, such as formaldehyde, chalk, and saccharin, cause a sour taste

---

*From the Middle English *jade*, a broken-down horse that is spiritless and crippled by fatigue.

when they mix with the citric and ascorbic acids of orange juice. Chewing the leaves of the asclepiad (a relative of the milkweed) makes one's ability to taste sweetness vanish. Sugar would taste bland and gritty. When Africans chew a berry they call "miraculous fruit," it becomes impossible to taste anything sour: lemons taste sweet, sour wine tastes sweet, rhubarb tastes sweet. Anything off-puttingly sour suddenly becomes delicious. A weak enough solution of salt tastes sweet to us, and some people salt melons to enhance the sweet flavor. Lead and beryllium salts can taste treacherously sweet, even though they're poisonous and we ought to be tasting them as bitter.

No two of us taste the same plum. Heredity allows some people to eat asparagus and pee fragrantly afterward (as Proust describes in *Remembrance of Things Past*), or eat artichokes and then taste any drink, even water, as sweet. Some people are more sensitive to bitter tastes than others and find saccharin appalling, while others guzzle diet sodas. Salt cravers have saltier saliva. Their mouths are accustomed to a higher sodium level, and foods must be saltier before they register as salty. Of course, everyone's saliva is different and distinctive, flavored by diet, whether or not they smoke, heredity, perhaps even mood.

How strange that we acquire tastes as we grow. Babies don't like olives, mustard, hot pepper, beer, fruits that make one pucker, or coffee. After all, coffee is bitter, a flavor from the forbidden and dangerous realm. To eat a pickle, one risks one's common sense, overrides the body's warning with sheer reason. *Calm down, it's not dangerous,* the brain says, it's novel and interesting, a change, an exhilaration.

Smell contributes grandly to taste. Without smell, wine would still dizzy and lull us, but much of its captivation would be gone. We often smell something before we taste it, and that's enough to make us salivate. Smell and taste share a common airshaft, like residents in a high rise who know which is curry, lasagna, or Cajun night for their neighbors. When something lingers in the mouth, we can smell it, and when we inhale a bitter substance—a nasal decongestant, for example—we often taste it as a brassiness at the back of the throat.

Smell hits us faster: It takes 25,000 times more molecules of cherry pie to taste it than to smell it. A head cold, by inhibiting smell, smothers taste.

We normally chew about a hundred times a minute. But, if we let something linger in our mouth, feel its texture, smell its bouquet, roll it around on the tongue, then chew it slowly so that we can hear its echoes, what we're really doing is savoring it, using several senses in a gustatory free-for-all. A food's flavor includes its texture, smell, temperature, color, and painfulness (as in spices), among many other features. Creatures of sound, we like some foods to titillate our hearing more than others. There's a gratifying crunch to a fresh carrot stick, a seductive sizzle to a broiling steak, a rumbling frenzy to soup coming to a boil, an arousing bunching and snapping to a bowl of breakfast cereal. "Food engineers," wizards of subtle persuasion, create products to assault as many of our senses as possible. Committees put a lot of thought into the design of fast foods. As David Bodanis points out with such good humor in *The Secret House,* potato chips are:

> an example of total destruction foods. The wild attack on the plastic wrap, the slashing and tearing you have to go through is exactly what the manufacturers wish. For the thing about crisp foods is that they're louder than non-crisp ones. . . . Destructo-packaging sets a favorable mood. . . . Crisp foods have to be loud in the upper register. They have to produce a high-frequency shattering; foods which generate low-frequency rumblings are crunchy, or slurpy but not crisp. . . .

Companies design potato chips to be too large to fit into the mouth, because in order to hear the high-frequency crackling you need to keep your mouth open. Chips are 80 percent air, and each time we bite one we break open the air-packed cells of the chip, making that noise we call "crispy." Bodanis asks:

> How to get sufficiently rigid cell walls to twang at these squeaking harmonics? Starch them. The starch granules in potatoes are identical to the starch in stiff shirt collars . . . whitewash . . . is . . . near identical in chemical composition. . . . All chips are soaked

in fat. . . . So it's a shrapnel of flying starch and fat that produces the conical air-pressure wave when our determined chip-muncher finally gets to finish her chomp.

These are high-tech potato chips, of course. The original potato chip was invented in 1853 by George Crum, a chef at Moon Lake Lodge in Saratoga Springs, New York, who became so angry when a guest demanded thinner and thinner French fries that he sliced them laughably thin (he thought) and fried them until they were varnish-brown. The guest loved them, envious fellow guests requested them, word spread, and ultimately Crum started up his own restaurant, which specialized in potato chips.

The mouth is what keeps the prison of our bodies sealed up tight. Nothing enters for help or harm without passing through the mouth, which is why it was such an early development in evolution. Every slug, insect, and higher animal has a mouth. Even one-celled animals like paramecia have mouths, and the mouth appears immediately in human embryos. The mouth is more than just the beginning of the long pipeline to the anus: It's the door to the body, the place where we greet the world, the parlor of great risk. We use our mouths for other things—language, if we're human; drilling tree bark if we're a woodpecker; sucking blood if we're a mosquito—but the mouth mainly holds the tongue, a thick mucous slab of muscle, wearing minute cleats as if it were an athlete.

THE ULTIMATE DINNER PARTY

Romans adored the voluptuous feel of food: the sting of pepper, the pleasure-pain of sweet-and-sour dishes, the smoldery sexiness of curries, the piquancy of delicate and rare animals, whose exotic lives they could contemplate as they devoured them, sauces that reminded them of the smells and tastes of lovemaking. It was a time of fabulous, fattening wealth and dangerous, killing poverty. The poor served the wealthy, and could be beaten for a careless word, destroyed for amusement. Among the wealthy, boredom visited like an impossible in-law, whom they devoted most of their lives to entertaining. Orgies and dinner parties were the main diversions, and the Romans amused themselves with the lavishness of a people completely untainted by annoying notions of guilt. In their culture, pleasure glistened as a good in itself, a positive achievement, nothing to repent. Epicurus spoke for a whole society when he asked:

> Is man then meant to spurn the gifts of Nature? Has he been born but to pluck the bitterest fruits? For whom do those flowers grow, that the gods make flourish at mere mortals' feet? . . . It is a way of pleasing Providence to give ourselves up to the various delights which she suggests to us; our very needs spring from her laws, and our desires from her inspirations.

Fighting the enemy, boredom, Romans staged all-night dinner parties and vied with one another in the creation of unusual and ingenious dishes. At one dinner a host served progressively smaller members of the food chain stuffed inside each another: Inside a calf, there was a pig, inside the pig a lamb, inside the lamb a chicken, inside the chicken a rabbit, inside the rabbit a dormouse, and so on. Another host served a variety of dishes that looked different but were all made from the same ingredient. Theme parties were popular, and might include a sort of treasure hunt, where guests who located the peacock brains or flamingo tongues received a prize. Mechanical devices might lower acrobats from the ceiling along with the next course, or send in a plate of lamprey milt on an eel-shaped trolley. Slaves brought garlands of flowers to drape over the diners, and rubbed their bodies with perfumed ungents to relax them. The floor might be knee-deep in rose petals. Course after course would appear, some with peppery sauces to spark the taste buds, others in velvety sauces to soothe them. Slaves blew exotic scents through pipes into the room, and sprinkled the diners with heavy, musky animal perfumes like civet and ambergris. Sometimes the food itself squirted saffron or rose water or some other delicacy into the diner's face, or birds flew out of it, or it turned out to be inedible (because it was pure gold). The Romans were devotees of what the Germans call *Schadenfreude,* taking exquisite pleasure in the misfortune of someone else. They loved to surround themselves with midgets, and handicapped and deformed people, who were made to perform sexually or cabaret-style at the parties. Caligula used to have gladiators get right up on the dinner table to fight, splashing the diners with blood and gore. Not all Romans were sadists, but numbers of the wealthy class and many of the emperors were, and they could own, torture, maltreat, or murder their slaves as much as they wished. At least one high-society Roman is recorded to have fattened his eels on the flesh of his slaves. Small wonder Christianity arose as a slave-class movement, emphasizing self-denial, restraint, the poor inheriting the earth, a rich and free life after death, and the ultimate punishment of the luxury-loving rich in the eternal tortures of hell. As Philippa Pullar observes in *Consuming Passions*, it was from this "class-consciousness and a pride in poverty and simplicity the hatred of the body was born. . . . All agreeable sensations were damned, all harmonies of taste and smell, sound, sight and feel, the candidate for heaven must resist them all. Pleasure was synonymous with guilt, it was synonymous with Hell. . . . 'Let your companions be women pale and thin with fasting,' instructed Jerome." Or, as Gibbon put it, "every sensation that is offensive to man was thought acceptable to God." So the denial of the senses became part of a Christian creed of salvation. The Shakers would later create their stark wooden benches, chairs, and simple boxes in such a mood, but what would they make now of the voluptuousness with which people enjoy Shaker pieces, not as a simple necessity but extravagantly, as art, as an expensive excess bought for the foyer or country house? The word "vicarious" hinges on "vicar," God's consul in the outlands, who lived like an island in life's racy current, delicate, exempt, and unflappable, while babies grew out of wedlock and bulls died, crops shriveled up like pokers or were flooded, and local duennas held musicales for vicar, matrons, and spicy young women (riper than the saintliest mettle could bear). No wonder they lived vicariously, giving pause, giving aid, and, sometimes, giving in to embolisms, dietary manias, and sin. Puritanism denounced spices as too sexually arousing; then the Quakers entered the scene, making all luxury taboo, and soon enough there were revolts against these revolts. Food has always been associated with cycles of sexuality, moral abandon, moral restraint, and a return to sexuality once again—but no one did so with as much flagrant gusto as the ancient Romans.

Quite possibly the Roman empire fell because of lead poisoning, which can cause miscarriages, infertility, a host of illnesses, and insanity. Lead suffused the Romans' lives—not only did their water pipes, cooking pots, and jars contain it, but also their cosmetics. But before it did poison them, they staged some of the wildest and most extravagant dinner parties ever known, where people dined lying down, two, three, or more to a couch. While saucy Roman poets like Catullus wrote rigorously sexy poems about affairs with either sex, Ovid wrote charming ones about his robust love of women, how they tormented his soul, and about the roller coaster of flirtation he observed at dinner parties. "Offered a sexless heaven," he wrote, "I'd say *no thank you,* women are such sweet hell." In one of his poems, he cautions his mistress that, since they've both been invited to the same dinner, he's bound to see her there with her husband. *Don't let him kiss you on the neck,* Ovid tells her, *it will drive me crazy.*

MACABRE MEALS

When the chic, sophisticated Romans conquered the wilds of Britain, their cuisine conquered, too. As Pullar has pointed out, the Anglo-Saxon words "cook" and "kitchen" derive from the Latin, so the Romans no doubt greatly raised the level of sophistication in both spheres. Medieval tastes were still Roman tastes (sweet and sour sauces, spicy, currylike dishes). It was the crusaders who developed a taste for the spices of the East—cinnamon, nutmeg, cardamom, mace, cloves, and rose attar—as they had for the perfumes, silks, dyes, ornate sexual practices, and other delicacies. The poor Britains lived in squalor and the rich lived in ostentation, holding magnificent feasts in honor of marriages and other celebrations. Many people have written that medieval cooks used a heavy hand with spices to mask the odor of their half-decayed meat, but ladling on the spices was a legacy from the Romans and the crusaders.

Some of the strangest culinary habits arose in England during the eighteenth century, when bored city dwellers became fascinated by sadism, sorcery, and a dungeons-and-skeletons sense of fun. The idea arose that torturing an animal made its meat healthier and better tasting and even though Pope, Lamb, and others wrote about the practice with disgust, people indulged in ghoulish preparations that turned their kitchens into charnel houses. They chopped up live fish, which they claimed made the flesh firmer; they tortured bulls before killing them, because they said the meat would otherwise be unhealthy; they tenderized pigs and calves by whipping them to death with knotted ropes; they hung poultry upside down and slowly bled them to death; they skinned living animals. Recipe openers from the era said such things as: "Take a red cock that is not too old and beat him to death. . . ." This was all sponsored by the peculiar notion that the taste of animal flesh could be improved if the poor thing were put through hell first. Dr. William Kitchiner, in *The Cook's Oracle*, cites a grotesque recipe, by a cook named Mizald, for preparing and eating a goose *while it is still alive:*

Take a goose, or a Duck, or some such lively creature pull off all her feathers, only the head and neck must be spared: then make a fire round about her, not too close to her, that the smoke do not choke her, and that the fire may not burn her too soon; not too far off, that she may not escape free: within the circle of the fire let there be set small cups and pots of water, wherein salt and honey are mingled; and let there be set also chargers full of sodden Apples, cut into small pieces in the dish. The Goose must be all larded, and basted over with butter: put then fire about her, but do not make too much haste, when as you see her begin to roast; for by walking about and flying here and there, being cooped in by the fire that stops her way out the unwearied Goose is kept in; she will fall to drink the water to quench her thirst, and cool her heart, and all her body, and the Apple sauce will make her dung and cleanse and empty her. And when she roasteth, and consumes inwardly, always wet her head and heart with a wet sponge; and when you see her giddy with running, and begin to stumble, her heart wants moisture, and she is roasted enough. Take her up and set her before your guests and she will cry as you cut off any part from her and will be almost eaten up before she be dead: it is mighty pleasant to behold!

THE HEART OF CRAVING

It's not to my taste, we say, by which we mean a hankering or preference, and it's amazing how individual taste can be—but only if survival is not at stake. When I worked on a cattle ranch in New Mexico, I used to eat in the cookhouse with the rest of the cowhands, most of whom were Mexican-Americans with little schooling and absolutely no education in nutrition. Their workdays were so arduous that their bodies took over for them, dictating what they needed to survive the physical labor and blinding heat of the day. Each morning, they would eat pure protein—as many as six eggs at once, with two glasses of whole milk, and bacon—for breakfast. Although they drank a lot of water and lemonade, they spurned coffee, tea, or other drinks with caffeine. They ate almost no desserts and very little sugar, but each meal included the hottest of hot peppers. Often they would spread them on bread to make a scalding jalapeño-pepper sandwich. At night they ate lightly, and the meal consisted mainly of carbohydrates. If asked, they would say simply that they ate what tasted good, what they liked to eat, but their taste in food had clearly evolved to fuel the rigors of their life.

This self-protective yen is also true on a larger scale: whole countries prefer cuisines that help them keep cool (in the Middle East), or sedated (in the tropics), or protect them against regional illnesses—as Pete Farb and George Armelagos say in their book which, like Pullars', is entitled *Consuming Passions,* "Ethiopian *chow,* consisting primarily of chili but containing up to fifteen other spices, has been shown to inhibit almost completely staphylococcus, salmonella, and other microorganisms." Hot peppers contain high amounts of beta carotene (converted by the body into vitamin A), which has antioxidant cancer-fighting properties, as well as capsaicin, which makes one sweat, lowering the body temperature. Consider the age-old English habit of drinking tea with milk: Tea contains a lot of tannin, which is toxic and can cause cancer, but milk protein reacts with the tannin in a protective way, preventing the body from absorbing it. Esophagal cancer is much higher in countries like Japan, where tea is drunk unadulterated, than it is in England, where people add a milk buffer to it. Farb and Armelagos describe some interesting additional national cravings:

> Peasants in Mexico prepare maize for making tortillas by soaking it in water in which they have previously dissolved particles of limestone, a practice which we certainly consider unusual. But . . . this preparation multiplies the calcium content to at least twenty times that in the original maize while possibly increasing the availability of certain amino acids—important because the peasants inhabit an environment where animal foods are scarce. . . . In places in Africa people eat fish wrapped in a banana leaf whose acidity dissolves the fish bones and thereby makes the calcium in them available; the French practice of cooking fish with sorrel has the same effect. Putrefied food . . . eaten in numerous societies . . . enhances the nutritive value . . . since the bacteria that cause putrefaction manufacture such vitamins as B[1]. . . .

There's no question that, at least for certain nutrients, if a person is in true need, some gustatory yen or body wisdom takes over. Patients with Addison's disease become ill because of a deficiency of the adrenal hormones. They've been known to crave salt with a vengeance, subconsciously medicating themselves. One way they do this is by eating large amounts of licorice, which contains glasorisic acid, a substance that causes sodium retention, and while doctors certainly don't prescribe it, they find that Addison's sufferers feel better if they eat a lot of licorice.

Some Quechua Indians of Peru subsist largely on potatoes, but because the growing season is so short, they're often forced to eat only partially ripened ones. Potatoes contain solanine, a bitter toxic alkaloid, but the Quechuas find that if they smear kaolin clay on the potatoes, it masks the bitterness and they don't get upset stomachs. The kaolin also detoxifies the alkaloids in the potatoes, making them simultaneously tastier and more nutritious.

It's odd to think of people eating dirt. Salt is the only rock we really seem to enjoy, but that's because we are small marine environments on the move, with salt in our blood, our urine, our flesh, our tears. However, you can still find clay for sale in some of the open-air markets in the southern United States. Pregnant women buy it. In

Africa, pregnant women occasionally eat termite mounds. It's thought that they're after calcium and certain other minerals missing from their diet. In Ghana, some villages support themselves by selling egg-shaped balls of clay, which are rich in potassium, magnesium, zinc, copper, calcium, iron, and other minerals. A pregnant woman's craving for dairy products makes good nutritional sense, because if the fetus doesn't get enough calcium, it will take it from the mother's bones and teeth. Most cultures have taboos for pregnant women, certain fish or fungi or spices they must not eat, but these are not the same as a woman's craving certain foods. The increased blood volume of a pregnant woman lowers her sodium level, and as a result she doesn't taste saltiness as easily as she did when she wasn't pregnant; she may crave really salty foods, like the legendary pickle. Among the many explanations for why pregnant women crave ice cream and other sweets, one of the most interesting modern theories is that they crave foods which produce the neurotransmitter serotonin, which they'll need to help withstand the pain of childbirth.

Some foods may stimulate endorphins—morphinelike painkillers produced by the brain—and give us a sense of comfort and calm. This is why, even though we know that salty foods, greasy foods, and candy and other sweets aren't good for us, we have a taste for them anyway. Neurobiologists suspect that endorphins and other neurochemicals control our hunger for certain kinds of foods. According to this thinking, when we eat sweets we flood our bodies with endorphins and feel tranquil. When people are under stress, and their need for endorphins goes up, they may crave a box of cookies. Since our hunger for fats, proteins, and carbohydrates is controlled by specific neurotransmitters, which can easily get out of balance, we need only binge to knock the neurotransmitters out of whack, which leads to further binging, further imbalances, and so on. In one experiment, depriving rats of their breakfast threw off their neurotransmitters and they gorged later in the day.

Are one's moods linked to food? Biochemist Judith Wurtman has published highly controversial findings about how food can affect our moods. She concludes that there are "carbohydrate cravers," who in reality are trying to raise their level of serotonin. When these levels are increased by drugs in controlled experiments, the carbohydrate cravers lose their cravings. Some scientists at the Monell Chemical Senses Institute and elsewhere dismiss her findings as being too tidy, too simple a version of how the body works, but I think some of it is persuasive. I never drink coffee after dinner, but I discovered accidentally over a period of years that I get to sleep better if I also don't eat protein late at night, only toast and jam or some other carbohydrates. On the other hand, around 3:30 in the afternoon, when my energies start to crash but I still have work to do, I'll be perked up by a jolt of protein, usually some cheese. My pattern gibes with Wurtman's experiments. The real power lunch, she suggests, revolves around an initial serving of protein, then a simple protein entreé and lightly cooked vegetables, with nothing richer than fruit for dessert, and no alcohol. Carbohydrates are sedating. When I meet someone for lunch and want to stay bright-eyed and bushy-tailed I order a high-protein appetizer like a shrimp cocktail or oysters on the half shell, or sliced mozzarella cheese with basil and tomatoes, and never nibble on the bread. A heap of pasta followed by chocolate mousse for dessert is what I'd really like, but I've found that it leaves me too listless to work. I disagree with Wurtman about why we crave chocolate, however—I don't think it's just a general cry for carbohydrate, but a craving for something more specific that chocolate provides.

Another researcher, one at the National Institute of Mental Health, found that people with Seasonal Affective Disorder (SAD), who become very depressed in winter, share a craving for carbohydrates at that time; this helps lift their mood. In yet another study, ex-smokers were found to crave carbohydrates. The link between carbohydrate craving, serotonin, and our drive to bring ourselves back into emotional balance seems undeniable. The brain is a chemical industry, and foods are highly complex chemicals. The extent to which eating one food or another may affect one's mood is really what's at issue.

Most people need about 15 percent of their food to be protein, and they automatically choose foods that will provide it, but scientists at the University of Toronto medical school discovered how much such a need can depend on genetics when they studied identical and fraternal twins. Identical twins, even though they were raised apart since birth, ate the same proportions of protein and carbohydrates, while fraternal twins didn't. So craving may, to some extent at least, be genetically determined. Hyperactive children often respond well to changes in their diet, as do those suffering from various disorders like Addison's disease or diabetes. But it's hard to say where memory stops and nutritional need or genetic fiat begins. We may crave sweets because we associate them with childhood rewards, or with being fed sweet liquids when we were babes-in-arms. Or we may crave them as a way to trigger the calm serotonin brings. Or both.

Most nutritionists, who are conservative, claim that there's no magic bullet and we should just try to eat as varied and well-balanced a diet as possible.* Under some circumstances, food can do more than change one's mood: It can kill. Raw liver used to be prescribed for pregnant women or those listless from iron deficiency, but now we know that liver collects the body's impurities and probably shouldn't be eaten at all. Polar bear liver is so high in vitamin A that it's toxic to humans. Alexander Pope and Henry I of England reportedly died from eating eels, which have poisonous filaments cooks might forget to remove. Balzac drank over fifty cups of coffee a day, and died from caffeine poisoning. Mushroom collectors run a steady risk of plucking the wrong fungus. Salmonella, which sounds so delicatessenlike and fresh, claims victims every year. Supposed aphrodisiacs have killed many victims, too. We don't think of plants as aggressive, but, since they can't run away from predators, they often devise extraordinary defense systems and potions, like strychnine, which protect them in the wild and sometimes appear on our plates.

----

*With one exception. Animals that are greatly underfed have longer life spans. Scientists aren't sure why—it may be the effect on the immune system, it may be the effect on metabolism, it may be something else entirely. And it's important that the animals not be undernourished, just fed a lot less than normal and given vitamin supplements. Studies are now beginning with primates, our closest relatives, but every other animal studied has shown longer life spans as a result of being skinnier.

THE PSYCHOPHARMACOLOGY
OF CHOCOLATE

What food do you crave? Ask the question with enough smoldering emphasis on the last word, and the answer is bound to be chocolate. It was first used by the Indians of Central and South America. The Aztecs called it *xocoatl* ("chocolate"), declared it a gift from their white-bearded god of wisdom and knowledge, Quetzalcoatl, and served it as a drink to members of the court—only rulers and soldiers could be trusted with the power it conveyed. The Toltecs honored the divine drink by staging rituals in which they sacrificed chocolate-colored dogs. Itzá human-sacrifice victims were sometimes given a mug of chocolate to sanctify their journey. What Hernán Cortés found surrounding Montezuma was a society of chocolate worshipers who liked to perk up their drink with chili peppers, pimiento, vanilla beans, or spices, and serve it frothing and honey-thick in gold cups. To cure dysentery, they added the ground-up bones of their ancestors. Montezuma's court drank two thousand pitchers of chocolate each day, and he himself enjoyed a chocolate ice made by pouring the drink over snow brought to him by runners from the mountains. Impressed by the opulence and restorative powers of chocolate, Cortés introduced it to Spain in the sixteenth century. It hit the consciousness of Europe like a drug cult. Charles V decided to mix it with sugar, and those who could afford it drank it thick and cold; they, too, occasionally added orange, vanilla, or various spices. Brillat-Savarin reports that "The Spanish ladies of the New World are madly addicted to chocolate, to such a point that, not content to drink it several times each day, they even have it served to them in church." Today, chocolate-zombies haunt the streets of every city, dreaming all day of that small plunge of chocolate waiting for them on the way home from work. In Vienna, the richest chocolate cakes are decorated with edible gold leaf. More than once, I've been seriously tempted to fly to Paris for the afternoon, just to go to Angelina, a restaurant on the rue de Rivoli where they melt a whole chocolate bar into each cup of hot chocolate. How many candy bars *don't* contain chocolate? Chocolate, which began as an upper-class drink, has become déclassé, trendy, cloaked in a tackiness it doesn't deserve. For example, an ad in *Chocolatier Magazine* offers a one-quarter-pound chocolate "replica of a 5¼-inch floppy disk." In fact, the company can provide an entire "computer work-station comprised of a chocolate terminal, chocolate computer keyboard, chocolate chip and chocolate byte." Their slogan is "Boots up into your mouth, not in your disk drive." One September weekend in 1984, the Fontainebleau Hotel in Miami offered a Chocolate Festival Weekend, with special rates, menus, and events. People could fingerpaint in chocolate syrup, attend lectures on chocolate, sample chocolates from an array of companies, learn cooking techniques, or watch a TV actor be dunked in six hundred gallons of chocolate syrup. Five thousand people attended. Chocolate festivals rage in cities all across America, and there are highly popular chocolate tours of Europe. In Manhattan last month I heard one woman, borrowing the jargon of junkies, say to another, "Want to do some chocolate?"

Because chocolate is such an emotional food, one we eat when we're blue, jilted, premenstrual, or generally in need of TLC, scientists have been studying its chemistry. In 1982, two psychopharmacologists, Dr. Michael Liebowitz and Dr. Donald Klein, proposed an explanation for why lovesick people pig out on chocolate. In the course of their work with intense, thrill-seeking women who go into post-thrill depressions, they discovered that they all had something remarkable in common—in their depressed phase, virtually all of them ate large amounts of chocolate. They speculated that the phenomenon might well be related to the brain chemical phenylethylamine (PEA), which makes us feel the roller coaster of passion we associate with falling in love, an amphetaminelike rush. But when the rush of love ends, and the brain stops producing PEA, we continue to crave its natural high, its emotional speed. Where can one find lots of this luscious, love-arousing PEA? In chocolate. So it's possible that some people eat chocolate because it reproduces the sense of well-being we enjoy when we're in love. A sly beau once arrived at my apartment with three Droste chocolate apples, and every wedge I ate over the next two weeks, melting lusciously in my mouth, filled me with amorous thoughts of him.

Not everyone agrees with the PEA hypothesis. The Chocolate Manufacturer's Association argues that:

> the PEA content of chocolate is extremely small, especially in comparison with that of some other commonly consumed foods. The standard serving size of three and a half ounces of smoked salami contains 6.7 mg of phenylethylamine; the same size serving of cheddar cheese contains 5.8 mg of phenylethlamine. The standard 1.5-ounce serving of chocolate (the size of the average chocolate bar) contains much less than 1 mg (.21 mg). Obviously, if Dr. Liebowitz's theory were true, people would be eating salami and cheese in far greater amounts than they are today.

And Dr. Liebowitz himself, in *The Chemistry of Love*, later asked of chocolate craving:

> Could this be an attempt to raise their PEA levels? The problem is that PEA present in food is normally quickly broken down by our bodies, so that it doesn't even reach the blood, let alone the brain. To test the effect of ingesting PEA, researchers at the National Institute of Mental Health ate pounds of chocolate, and then measured the PEA levels in their urine for the next few days; the PEA levels didn't budge.

As a thoroughgoing chocoholic, I should say that I do indeed eat a lot of cheese. Smoked salami is too unhealthy for me even to consider; the Cancer Society has suggested that people should not eat foods that are smoked or contain nitrites. So, it's entirely possible that cheese fills some of my PEA need. What else do chocoholics eat? In other words, what is the total consumption of PEA from all sources? Chocolate may be a more appealing, even if smaller, source of PEA because of its other associations with luxury and reward. The NIMH study tested average people, but suppose people who crave chocolate aren't average? Isn't that the idea? Liebowitz now says that PEA may break down too fast to affect the brain. We still know very little about the arcane ways in which some drugs do this, not enough to completely dismiss chocolate's link with PEA.

Wurtman and others argue that we crave chocolate because it's a carbohydrate, which, like other carbohydrates, prompts the pancreas to make insulin, which ultimately leads to an increase in that neurotransmitter of calm, serotonin. If this were true, a plate of pasta, or potatoes, or bread would be equally satisfying. Chocolate also contains theobromine ("food of the gods"), a mild, caffeinelike substance, so, for the sake of argument, let's say it's just the serotonin and the relative of caffeine we crave, a calm stimulation, a culinary oxymoron few foods provide.* It might even explain why some women crave chocolate when they're due to menstruate, since women who suffer PMS have been found to have lower levels of serotonin, and premenstrual women in general eat 30 percent more carbohydrates than they do at other times of the month. But if it were as simple as that, a doughnut and a cup of coffee would do the trick. Furthermore, there's a world of difference between people who enjoy chocolate, women who crave chocolate only at certain times of the month, and serious chocoholics. Chocoholics don't crave potato chips and pasta; they crave chocolate. Substitutes in any combination won't do. Only the chocoholic in a household fresh out of chocolate, on a snowy night when the roads are impassable, knows how specific that craving can be. I'm not sure why some people crave chocolate, but I am convinced that it's a specific need, and therefore the key to solving a specific chemical mystery to which we'll one day find the solution.

The Four Seasons restaurant in Manhattan serves a chocolate bombe that's the explosive epitome of chocolate desserts, two slices of which (the standard serving) few people are able to finish because it's so piquantly rich. On the waterfront in St. Louis I once had a mousse called "Chocolate Suicide," which was drug-level chocolate. I felt as if my brain had been hung up in a smokehouse. I can still remember the first time I had Godiva chocolates at a friend's house; they were Godivas from the original factory in Brussels, with a perfect sheen, a twirling aroma, heady but not jarring, and a way of

---

*In a one-and-a-half-ounce milk-chocolate bar, there are about nine milligrams of caffeine (which the plant may use as an insecticide); a five-ounce cup of brewed coffee has about 115 milligrams; a twelve-ounce cola drink between thirty-two and sixty-five.

delicately melting on the tongue. One of the reasons why chocolates are superb in Belgium, Vienna, Paris, and some of our American cities is that chocolate candy is in considerable part a dairy product. The chocolate flavor may come from the plant, but the silken, melting delight comes from the milk, cream, and butter, which must be fresh. The people who create designer chocolates have learned that their confections must provide just the right melting sensation, and feel quintessentially creamy and luscious, with no grittiness or aftertaste, for people to be thoroughly wowed by them. In George Orwell's *1984*, sex is forbidden and chocolate is "dull-brown crumbly stuff that tasted . . . like the smoke of a rubbish fire." Just before Julia and Winston risk making love, they eat real, full-bodied "dark and shiny" chocolate. Their amorous feast had its precedents. Montezuma drank an extra cup of chocolate before he went to visit his women's quarters. Glamorous movie stars like Jean Harlow used to be shown eating boxes of chocolates. M. F. K. Fisher, the diva of gastronomy, once confided that her mother's doctor prescribed chocolate as a cure for debilitating lovesickness. On the other hand, Aztec women were forbidden chocolate; what secret terror was it thought to unleash in them?

IN PRAISE OF VANILLA

Craving vanilla, I start the bathwater gushing, and unscrew the lid of a heavy glass jar of Ann Steeger of Paris's Bain Crème, *senteur vanille*. A wallop of potent vanilla hits my nose as I reach into the lotion, let it seep through my fingers, and carry a handful to the faucet. Fragrant bubbles fill the tub. A large bar of vanilla bath soap, sitting in an antique porcelain dish, acts as an aromatic beacon. While I steep in waves of vanilla, a friend brings me a vanilla cream seltzer, followed by a custard made with vanilla beans that have come all the way from Madagascar. Brown flecks float through the creamy yellow curds. Though I could have chosen beans from the Seychelles, Tahiti, Polynesia, Uganda, Mexico, the Tonga Islands, Java, Indonesia, the Comoro Islands, and other places, I like the long, sensuous shape of the Madagascar vanilla bean, and its dark, rich, pliable coat, which looks like carefully combed tresses or the pelt of a small aquatic animal. Some connoisseurs prefer the shorter Tahitian bean, which is fatter and moister (even though it has less vanillin and the moistness is only water, not flavorful oils), or the smoky flavor of beans from Java (wood fires do some of the curing), or the maltier flavor of those from the Comoros.

Most of the world's real vanilla comes from islands in the Indian Ocean (Madagascar, Réunion, Comoros), which produce a thousand tons of vanilla beans every year. But we rarely taste the real thing. The vanilla flavoring we buy in the spice section of grocery stores, the vanilla we find in most of our ice creams, cakes, yogurts, and other foods, as well as in shampoos and perfumes, is an artificial flavor created in laboratories and mixed with alcohol and other ingredients. Marshall McLuhan once warned us that we were drifting so far away from the real taste of life that we had begun to *prefer* artificiality, and were becoming content with eating the menu descriptions rather than the food. Most people have used the medicinal-smelling artificial vanilla flavoring for so long that they have no idea what real vanilla extract tastes and smells like. Real vanilla, with its complex veils of aroma and jiggling flavors, makes the synthetic seem a poor parody. Vanillin isn't the only flavor in genuine vanilla, but it's the one synthetically produced (originally from clove oil, coal tar, and other unlikely substances, but now mainly from the sulfite by-products of paper manufacturing). Indeed, the world's largest producer of synthetic vanillin is the Ontario Paper Company! Real vanilla varies along a spectrum from sweet and dusty to damp and loamlike, depending on the variety of bean, its freshness, its home country, how and for how long it was cured and in what temper of sun.

When a vanilla bean lies like a Hindu rope on the counter, or sits in a cup of coffee, its aroma gives the room a kind of stature, the smell of an exotic crossroads where outlandish foods aren't the only mysteries. In Istanbul in the 1970s, my mother and I once ate Turkish pastries redolent with vanilla, glazed in caramel sugar with delicate filaments of syrup on top. It was only later that day, when we strolled through the bazaar with two handsome university students my mother had bumped into, that we realized what we had eaten with such relish. On a long brass platter sat the kind of pastries we had eaten, buzzed over by hundreds of sugar-delirious bees, whose feet stuck in the syrup; desperately, one by one, they flew away, leaving their legs behind. "Bee legs!" my mother had screamed, as her face curdled. "We ate *bee legs!*" Our companions spoke little English and we spoke no Turkish, so they probably thought it odd that American women became so excitable in the presence of pastry. They offered to buy us some, which upset my mother even more.

Walk through a kitchen where vanilla beans are basking in a loud conundrum of smell, and you'll make some savoring murmur without realizing it. The truth about vanilla is that it's as much a smell as a taste. Saturate your nose with glistening, soulful vanilla, and you can *taste* it. It's not like walking through a sweetshop, but more subterranean and wild. Surely this is the unruly beast itself, the raw vanilla that's clawing your senses. But no. The vanilla beans we treasure aren't delectable the way we find them in the jungle. Of all the foods grown domestically in the world, vanilla requires the most labor: Long, tedious hours of hand tending bring the vanilla orchids to fruit and then the fruit to lusciousness. Vanilla comes from the string-bean-like pod of a climbing orchid, whose greenish-white flowers bloom briefly and are without fragrance. Since the blossoms last only one day, they must be hand-pollinated exactly on schedule. The beans mature six weeks after fertilization, but cannot be picked for some months longer. When a bean turns perfectly ripe, the pickers plunge it into boiling water to stop the ripening; they dry and process it, using blankets, ovens, racks, and sweating boxes; and slowly cure it in the sun for six to nine months. The glorious scent and taste don't adorn the growing plant. It's only as the beans ferment to wrinkled, crackly brown pods that the white dots of vanillin crystallize mellowly on their outsides and that famous robust aroma starts to saturate the air.

It was in 1518 that Cortés first noticed the Aztecs flavoring their chocolate with ground-up vanilla pods, which they called *tlilxochitl* ("black flower") and prized so highly that Montezuma drank an infusion of it as a royal balm and demanded vanilla beans in tribute from his subjects. The Spaniards called the bean *vainilla* ("small sheath"), from the Latin *vagina*—the bean's elongated shape, with a slit at the top, must have reminded the lonesome Spaniards of what they were missing. There would have been many boisterous jokes about Montezuma stirring his chocolate with a little vagina.* Cortés valued vanilla enough to carry bags of it back to Europe, along with the Aztecs' gold, silver, jewels, and chocolate. A passion for vanilla, especially in combination with chocolate, raged in Europe, where it was prized as an aphrodisiac. Thomas Jefferson's letters include an appeal to a Parisian friend to send him some vanilla beans, for which he had developed a taste during his tenure as the U.S. minister to France, and which he couldn't find in American apothecary shops.

Precious and desirable as vanilla was, no one could figure out how to grow it outside Mexico. The problem was typical of the delicate ecosystem in the rain forest, and a good example of how fragile all that lush green abandon really is, but no one realized it. Though insects, birds, and bats pollinate most plants in the tropics, the vanilla orchid is pollinated by only one type of bee, the tiny Melipone. In 1836, a Belgian figured out the vanilla orchid's secret sex life when he caught sight of the Melipone bumbling about its work. Then the French devised a method of hand-pollinating the orchids and started plantations on their Indian Ocean islands, as well as in the East and West Indies. The Dutch carried vanilla to Indonesia, and the British to India. "Tincture of vanilla" didn't appear in the United States until the 1800s, but when it did, it appealed to the American impatience and aversion to fuss, that sprint through life whose byword is *convenience*. Europeans used the vanilla bean, luxuriating in its textures, tastes, and aromas, but we preferred it reduced and already bottled. By the nineteenth century, demand flourished, vanilla became synthesized, and the world floated on a mantle of cheap flavoring. Vanilla now appears as an ingredient in

---

*Randy workmen and explorers are responsible for a lot of interesting etymology. Consider the word "gasket," which comes from the Old French *garcette*, a little girl with her hymen still intact.

most baked goods and in many perfumes, cleaning products, and even toys, and has insinuated itself into the cuisine of far-flung peoples, conquering their palates. Only saffron is a more expensive spice.

When I finally emerge from the tub into which I climbed at the beginning of this discussion, I apply Ann Steeger's vanilla body veil, which smells edible and thick as smoke. Then Jean Laporte's Vanilla perfume, vanilla with a bitter sting. The inside of a vanilla bean contains a figlike marrow, and if I were to scrape some out, I could prepare spicy vanilla bisque for dinner, followed by chicken in a vanilla glaze, salad with vanilla vinaigrette, vanilla ice cream with a sauce of chestnuts in vanilla marinade, followed by warm brandy flavored with chopped vanilla pod, and then, in a divine vanilla stupor, seep into bed and fall into a heavy orchidlike sleep.*

THE TRUTH ABOUT TRUFFLES

"The world's homeliest vegetable," it's been called, but also "divinely sensual" and possessing "the most decadent flavor in the world." As expensive as caviar, truffles sell for over $500 a pound in Manhattan these days, which makes it the most expensive vegetable on earth. Or, rather, under earth. Truffle barons must depend on luck and insight. A truffle may be either black (*melanosporum*) or white (*magnata*), and can be cooked whole, though people usually shave raw slivers of it over pasta, eggs, or other culinary canvases. For 2,000 years it's been offered as an aphrodisiac, prized by Balzac, Huysmans, Colette, and other voluptuous literary sorts for its presumed ability to make one's loins smolder like those of randy lions. When Brillat-Savarin describes the dining habits of the duke of

---

*To make real vanilla extract: Split a vanilla bean lengthwise, set in a glass jar, cover with ¾ cup vodka. Cover and let steep for at least six weeks. As you use the extract, add more vodka; the bean will stay redolent and continue oozing flavor for some time. Add a teaspoon of vanilla extract to French toast batter to transmogrify it into the New Orleans version called "lost bread." Vanilla sugar tastes wonderful in coffee: Split one vanilla bean from top to bottom and cut into pieces; mix with two cups of sugar; cover; let stand for six weeks. The longer the vanilla stands, the more intense the flavor.

Orleans, he gets so excited about the truffles that he uses three exclamation points:

> Truffled turkeys!!! Their reputation mounts almost as fast as their cost! They are lucky stars, whose very appearance makes gourmands of every category twinkle, gleam, and caper with pleasure.

One writer describes the smell of truffles as "the muskiness of a rumpled bed after an afternoon of love in the tropics." The Greeks believed truffles were the outcome of thunder, reversed somehow and turned to root in the ground. Périgord, in southwest France, produces black truffles that ooze a luscious perfume and are prized as the ne plus ultra of truffles, essential black sequins in the famous Périgord goose-liver pâté. The best white truffles come from the Piedmont region, near Alba in Italy. Napoleon is supposed to have conceived "his only legitimate son after devouring a truffled turkey," and women throughout history have fed their male companions truffles to rouse their desire. Some truffle dealers use trained dogs to locate the truffles, which tend to grow close to the roots of some lindens, scrub oaks, and hazelnut trees; but sows are still the preferred truffle hunters, as they have been for centuries. Turn a sow loose in a field where there are truffles, and she'll sniff like a bloodhound and then dig with manic passion. What is the sow's obsession with truffles? German researchers at the Technical University of Munich and the Lübeck School of Medicine have discovered that truffles contain twice as much androstenol, a male pig hormone, as would normally appear in a male pig. And boar pheromone is chemically very close to the human male hormone, which may be why we find truffles arousing, too. Experiments have shown that if a little bit of androstenol is sprayed into a room where women are looking at pictures of men, they'll report that the men are more attractive.

For the truffle farmer and his sow, walking above a subterranean orchard of truffles, it must be hysterically funny and sad. Here this beautiful, healthy sow smells the sexiest boar she's ever encountered in her life, only for some reason he seems to be underground. This drives her wild and she digs frantically, only to turn up a strange, lumpy, splotched mushroom. Then she smells another supermacho boar only a few feet away—also buried underground—and dives in, trying desperately to dig up that one. It must make her berserk with desire and frustration. Finally, the truffle farmer gathers the mushrooms, puts them in his sack, and drags his sow back home, though behind her the whole orchard vibrates with the rich aromatic lust of handsome boars, every one of them panting for her, but invisible!

GINGER, AND OTHER MEDICINES

On a voyage to the Antarctic in tempestuous waters, I become seasick and crawl into my cabin for a rest. But my cabin is aft and high on the cruise ship, and rolls far around the moment arm of the ship, then leaps up with each wave and crashes down, rolls and leaps again, occasionally throwing in a shimmy for good measure. Unscrewing a small jar of stubby brown knots, I roll one out, place it in my mouth, suck on it to soften it, then methodically begin to chew as a pleasant searing oozes over my tongue. Ginger has a long history of medicinal use in China, where they drink ginger tea for colds, flu, and other ailments. Chinese fishermen chew on ginger root to prevent seasickness.

Over the past few years, researchers around the world have been testing ginger's folkloric reputation, and have found this knotty root to live up to its legend. Researchers in Japan discovered that ginger is indeed a good cough suppressant; furthermore, it acts as an analgesic, lowers temperature, stimulates the immune system, and calms the heart in general, while at the same time strengthening the beating of the atrium, just as digitalis does. Nigerian scientists found that it acts as an antioxidant, and can kill salmonella. In California, scientists discovered that it works as a potent meat tenderizer and preserver. In a joint study at Brigham Young University in Utah and Mount Union College in Ohio, researchers learned that ginger acts better than Dramamine to keep motion sickness at bay. In Denmark, experiments showed that ginger keeps the blood from forming clots. In India, they discovered that ginger lowers cholesterol.

With all the edicts about what to eat when and what to avoid, it sometimes feels as if we're medicating ourselves rather than dining. Aluminum pots are out, since microscopic particles of aluminum can get into the food, and aluminum has been implicated in Alzheimer's disease. Butter, cream, and saturated fats are out, since they can lead to heart disease. Fiber is in, since it can help prevent rectal cancer, but not too much fiber, which can be equally damaging. Green, leafy vegetables are in for their antioxidant effect—but not if you're on a blood thinner, because they contain vitamin K, which clots blood. Fish oils are in, because they're important for the heart, but fish are often found to contain pollutants. Fresh fruit is important for its vitamin C, fiber, and other elements, although frequently sprayed with insecticide that's carcinogenic. Beef is out because of its high fat content, which has been implicated in everything from polyps to breast cancer, and, anyway, grilling meat produces carcinogens. Poultry is often fed hormones that aren't good for us, and frequently contains salmonella. Shellfish, as a light low-fat source of protein, is all right, but one must be careful to order oysters that haven't come from polluted harbors; and is it really safe to eat lobster and shrimp, both high in cholesterol, which are scavengers, i.e., creatures who eat the putrid remains of other creatures? In this morass of paradoxes, how on earth can one guiltlessly consider taste?

As a culture, we are mesmerized by the idea of the medicinal quality of food, swearing by yogurt, bean curd, carrot juice, ginseng root, raw honey, and many other items as they drift in and out of fashion. We forget that, in our not-too-distant past, the landscape was our pharmacy; it still is for many native peoples, as well as for the most sophisticated drug companies, who continue to send people into the rain forests to gather leaves for all manner of drugs. "Tell me what you eat, and I shall tell you what you are," Brillat-Savarin once said, but we understand his maxim in a broader sense than he did, picturing all the vitamins that heal, proteins that strengthen, fibers that scour and protect, carbohydrates that calm, sugars that energize. Children of the industrial age, we still think of eating as fueling our bodies, stoking the tiny furnace in each cell. We picture our body as a factory, and sometimes even use that word when we talk about its processes. Many of our creations resemble us. For a while, neurologists railed against comparing the brain to a computer, because it seemed terrifyingly automatic, amoral, and mechanistic. Now the computer simile is back in vogue, because the similarities are so obvious as to be undeniable. The brain is the computer; religion, prejudice, bias, and so forth are all software. The neurologists haven't become more coldblooded all of a sudden; computers have just become more familiar and less frightening entities. Yes, we say, brains that needed to store more information than they could hold invented artificial brains that merely reproduced the filing system that was familiar to them. No surprise in that. When we wished to create energy outside of our bodies, we also copied the only model we knew: You put fuel into something and it empowers it for a while, excretes wastes, and needs to be fed again to do more work. What great analogizers we are. It's part of our greatest charm as a species that we can look at the footprint of an elephant in the dried mud beside a waterhole, see how its steep sides trap water, and say: I could use one of those to carry liquids. In *Henry IV, Part II,* Shakespeare has Falstaff say that the body serves as our model of society as well, that the body has its own politics and classes. But analogies can run both ways, like an alternating current. Not only do we create mechanical powerhouses on the principle of the body, we eat candy bars called Powerhouse to power our body. And, whatever our age, we all eat some foods we secretly detest, because we suspect they're therapeutic. We prescribe foods: "Eat your broccoli," we insist, thinking of its gifts of vitamins and fiber, not that it looks like a small forest floating in the pot. "It's good for you."

HOW TO MAKE MOOSE SOUP IN A HOLE IN THE GROUND, OR DINE IN SPACE

In a small bedside bookcase, I often keep bare-bones survival texts like *A Pilot's Survival Manual,* from which one learns the correct side of a nomad's tent to enter after crash-landing in the Gobi

Desert, or Bradford Angier's *How To Stay Alive in the Woods,* with this recipe for moose soup made in a hole in the ground:

> You've just killed a moose. Hungry, you've a hankering for nothing quite as much as some hot soup, flavored perhaps with wild leeks whose flat leaves you see wavering nearby. Why not take the sharp end of a dead limb and scoop a small hole in the ground? Why not line this concavity with a chunk of fresh hide? Then after adding the water and·other ingredients, why not let a few hot clean stones do your cooking while you finish dressing out the animal?

Indeed, why not? I particularly like the recipe's opening: *You've just killed a moose.* It reminds me of a recipe I once read for stir-fried dog, which began: *First clean and eviscerate a healthy puppy.* If, like me, you try not to eat mammals unless pressed by an unknowing host or necessity (a knowing host), neither dish will make your mouth water. But I like the idea of quietly brewing moose soup in a mossy pit. This book assumes that though clothed, armed, and equipped with a compass, one may have forgotten matches. Cooking, while not essential to survival, certainly makes it easier, so there are many plans for starting a fire with water (used as a magnifier), watches (hold "the crystals from two watches or pocket compasses of about the same size back to back . . ."), a drill made out of a bow, sparking a hunting knife against flint and other paraphernalia, including a gun.*

Think what the survival manuals for space travel will include! Much of the pleasure of taste is smell; we can smell something only when it evaporates. So, I imagine there are fewer scents in weightlessness. And that would mean food wouldn't taste as good. Nonetheless, competition is keen to cater the Soviet and American space shuttles. One likely supplier for the next Soviet shuttle is Belème, a company jointly owned by a French astronaut, a biologist who studies weightlessness, and the chef and owner of L'Espérance, a three-star Miche-

---

*"Pry the bullet from the cartridge, first loosening the case if you want by laying it on a log and tapping the neck all around with the back of your knife. . . . Have the campfire laid with a good bed of tinder beneath. Pour some of the powder over this tinder. Stuff a small bit of dry frayed cloth into the remains of the load. Fire the weapon straight up into the air. The rag, if it is not already burning when it falls nearby, should be smoldering sufficiently so that when pressed into the tinder it can be quickly blown into flame."

lin restaurant near Paris. The orbital menu would include such *haute* delicacies as artichoke chips and *poulet à la Dijonnaise,* presented in tubes and cans. Belème already supplies polar and desert explorers, mountain climbers, racing-car drivers, and other gastronomically aware adventurers with gourmet foods appropriate to the environment they'll be in. When we think of cuisines, we picture steaming plates of curry, crawfish, peanut soup, chili, fettuccine, or some other savory dialect. But there is also, in its infancy, a space cuisine. I've eaten NASA's freeze-dried space peaches, which taste like sweetly citric wasp's nest, and read astronauts' accounts of other foods; space cuisine is nothing to write home about. But wonder flavors things better than any condiment, so for short hauls freeze-dried fare may do just fine, until space travel is no stranger than a stroll along the Rialto in Venice, and we dare to dine al fresco at a cozy little spot whose menu offers moon on the half shell and a side order of stars.

ET FUGU, BRUTE?
FOOD AS THRILL-SEEKING

A nation of sensation-addicts might dine as chic urbanites do, on rhubarb and raspberry tortes, smoked lobster, and hibiscus-wrapped monkfish, wiped with raspberry butter, baked in a clay oven, and then elevated briefly in mesquite smoke. When I was in college, I didn't eat goldfish or cram into Volkswagens, or chug whole bottles of vodka, but others did, in a neo–Roaring Twenties ennui. Shocking the bourgeoisie has always been the unstated encyclical of college students and artists, and sometimes that includes grossing out society in a display of bizarre eating habits. One of the classic *Monty Python's Flying Circus* sketches shows a chocolate manufacturer being cross-examined by policemen for selling chocolate-covered baby frogs, bones and all ("without the bones, they wouldn't be crunchy!" he whines), as well as insects, and other taboo animals sure to appal western taste buds. I've met field scientists of many persuasions who have eaten native foods like grasshoppers, leeches, or bats stewed in coconut milk, in part to be mannerly, in part out of curiosity, and I think in part to provide a good anecdote when they returned to the States. However, these are just nutritious foods that fall beyond our usual sphere of habit and custom.

We don't always eat foods for their taste, but sometimes for their feel. I once ate a popular duck dish in Amazonian Brazil, *pato no tucupí* (Portuguese for *pato*, "duck" + *no*, "within" + *tucupí*, "extracted juice of manioc") whose main attraction is that it's anesthetic: It makes your mouth as tingly numb as Benzedrine. The numbing ingredient is *jambu* (in Latin, *Spilanthes*), a yellow daisy that grows throughout Brazil and is sometimes used as a cold remedy. The effect was startling—it was as if my lips and whole mouth were vibrating. But many cultures have physically startling foods. I adore hot peppers and other spicy foods, ones that sandblast the mouth. We say "taste," when we describe such a food to someone else, but what we're really talking about is a combination of touch, taste, and the absence of discomfort when the deadening or sandblasting finally stops. The thinnest line divides Szechwan hot-pepper sauce from being thrilling (causing your lips to tingle even after the meal is over), and being sulfurically hot enough to cause a gag response as you eat it.* A less extreme example is our liking for crunchy or crisp foods, like carrots, which have little taste but lots of noise and mouth action. One of the most successful foods on earth is Coca-Cola, a combination of intense sweetness, caffeine, and a prickly feeling against the nose that we find refreshing. It was first marketed as a mouthwash in 1888, and at that time contained cocaine, a serious refresher—an ingredient that was dropped in 1903. It is still flavored with extract of coca leaves, but minus the cocaine. Coffee, tea, tobacco, and other stimulants all came into use in the western world in the sixteenth and seventeenth centuries, and quickly percolated around Europe. Fashionable and addictive, they offered diners a real nervous-system jolt, either of narcotic calm or caffeine rush, and, unlike normal foods, they could be taken in doses, depending on how high one wished to get or how addicted one already was.

---

*Water won't work as an antidote because it doesn't mix with oil, the binding in Chinese food; plain rice is the best remedy.

In Japan, specially licensed chefs prepare the rarest sashimi delicacy: the white flesh of the puffer fish, served raw and arranged in elaborate floral patterns on a platter. Diners pay large sums of money for the carefully prepared dish, which has a light, faintly sweet taste, like raw pompano. It had better be carefully prepared, because, unlike pompano, puffer fish is ferociously poisonous. You wouldn't think a puffer fish would need such chemical armor, since its main form of defense is to swallow great gulps of water and become so bloated it is too large for most predators to swallow. And yet its skin, ovaries, liver, and intestines contain tetrodotoxin, one of the most poisonous chemicals in the world, hundreds of times more lethal than strychnine or cyanide. A shred small enough to fit under one's fingernail could kill an entire family. Unless the poison is completely removed by a deft, experienced chef, the diner will die midmeal. That's the appeal of the dish: eating the possibility of death, a fright your lips spell out as you dine. Yet preparing it is a traditional art form in Japan, with widespread aficionados. The most highly respected *fugu* chefs are the ones who manage to leave in the barest touch of the poison, just enough for the diner's lips to tingle from his brush with mortality but not enough to actually kill him. Of course, a certain number of diners do die every year from eating *fugu*, but that doesn't stop intrepid *fugu*-fanciers. The ultimate *fugu* connoisseur orders *chiri*, puffer flesh lightly cooked in a broth made of the poisonous livers and intestines. It's not that diners don't understand the bizarre danger of puffer-fish toxin. Ancient Egyptian, Chinese, Japanese, and other cultures all describe *fugu* poisoning in excruciating detail: It first produces dizziness, numbness of the mouth and lips, breathing trouble, cramps, blue lips, a desperate itchiness as of insects crawling all over one's body, vomiting, dilated pupils, and then a zombielike sleep, really a kind of neurological paralysis during which the victims are often aware of what's going on around them, and from which they die. But sometimes they wake. If a Japanese man or woman dies of *fugu* poison, the family waits a few days before burying them, just in case they wake up. Every now and then someone poisoned by *fugu* is nearly buried alive, coming to at the last moment to describe in horrifying detail their own funeral and burial, during which, although they desperately tried to cry out or signal that they were still alive, they simply couldn't move.

Though it has a certain Russian-roulette quality to it, eating *fugu* is considered a highly aesthetic experience. That makes one wonder about the condition that we, in chauvinistic shorthand, refer to as "human." Creatures who will one day vanish from the earth in that ultimate subtraction of sensuality that we call death, we spend our lives courting death, fomenting wars, watching sickening horror movies in which maniacs slash and torture their victims, hurrying our own deaths in fast cars, cigarette smoking, suicide. Death obsesses us, as well it might, but our response to it is so strange. Faced with tornadoes chewing up homes, with dust storms ruining crops, with floods and earthquakes swallowing up whole cities, with ghostly diseases that gnaw at one's bone marrow, cripple, or craze—rampant miseries that need no special bidding, but come freely, giving their horror like alms—you'd think human beings would hold out against the forces of Nature, combine their efforts and become allies, not create devastations of their own, not add to one another's miseries. Death does such fine work without us. How strange that people, whole countries sometimes, wish to be its willing accomplices.

Our horror films say so much about us and our food obsessions. I don't mean the ones in which maniacal men carting chain saws and razors punish single women for living alone or taking jobs—although those are certainly alarming. I don't mean ghost stories, in which we exhale loudly as order falls from chaos in the closing scenes. And I don't mean scary whodunits, at the end of which the universe seems temporarily less random, violent, and inexplicable. Our real passion, by far, is for the juiciest of horror films in which vile, loathsome beasts, gifted with ferocious strength and cunning, stalk human beings and eat them. It doesn't matter much if the beast is a fast-living "Killer Shrew" or a sullen "Cat People" or an abstract "Wolfen" or a nameless, acid-drooling "Alien." The pattern is always the same. They dominate the genre. We are greedy for their brand of terror.

The plain truth is that we don't seem to have gotten used to being at the top of our food chain. It must bother us a great deal, or we wouldn't keep making movies, generation after generation, with exactly the same scare tactics: The tables are turned and we become fodder. All right, so we may be comfortable at the top of the chain as we walk around Manhattan, but suppose—oh, ultimate horror!— that on other planets *we're* at the bottom of *their* food chain? Then you have the diabolically scary "Aliens," who capture human beings, use them as hosts for their maggotlike young, and actually hang them up on slime gallows in a pantry.

We rush obsessively to movie theaters, sit in the cavelike dark, and confront the horror. We make contact with the beasts and live through it. The next week, or the next summer, we'll do it all over again. And, on the way home, we keep listening for the sound of claws on the pavement, a supernatural panting, a vampiric flutter. We spent our formative years as a technologyless species scared with good reason about lions and bears and snakes and sharks and wolves that could, and frequently did, pursue us. You'd think we'd have gotten over that by now. One look at the cozy slabs of cow in a supermarket case, neatly cut, inked, and wrapped, should tell us to relax. But civilization is a more recent phenomenon than we like to think. Are horror films our version of the magic drawings on cave walls that our ancestors confronted? Are we still confronting them?

Fugu might not seem to have much to do with nuclear disarmament or world peace, but it's a small indicator of our psyches. We find the threat of death arousing. Not all of us, and not all the time. But enough do often enough to keep the rest of us peace-loving sorts on our toes when we'd rather be sitting down calmly to a sumptuous meal with friends.

BEAUTY AND THE BEASTS

In Jean Cocteau's extraordinary film version of the classic fairy tale "Beauty and the Beast," a sensitive beast lives in a magical castle, the walls and furnishings of which are all psychosensitive. On the back of the Beast's chair, in Latin, runs the motto: *All men are beasts when they don't have love.* Every evening, the literate, humane beast must go out hunting for his dinner, chase down a deer and feed on its steaming flesh, or die of starvation. Afterward, he suffers the most bitter anguish, and his whole body involuntarily begins to smoke. The unstated horror of our species reveals itself in that moment. Like the sensitive Beast, we must kill other forms of life in order to live. We must steal their lives, sometimes causing them great pain. Every one of us performs or tacitly approves of small transactions with torture, death, and butchery each day. The cave paintings reflected the reverence and the love the hunter felt for his prey. In our hearts, we know that life loves life. Yet we feast on some of the other life-forms with which we share our planet; we kill to live. Taste is what carries us across that rocky moral terrain, what makes the horror palatable, and the paradox we could not defend by reason melts into a jungle of sweet temptations.

Hearing

I was all ear,
And took in strains that might create a soul
Under the ribs of Death.

John Milton, "Comus"

THE HEARING HEART

In Arabic, absurdity is not being able to hear. A "surd" is a mathematical impossibility, the core of the word "absurdity," which we get from the Latin *surdus*, "deaf or mute," which is a translation from the Arabic *jadr asamm*, a "deaf root," which in turn is a translation from the Greek *alogos*, "speechless or irrational." The assumption hidden in this etymological nest of spiders is that the world will still make sense to someone who is blind or armless or minus a nose. But if you lose your sense of hearing, a crucial thread dissolves and you lose track of life's logic. You become cut off from the daily commerce of the world, as if you were a root buried beneath the soil. Despite Keats's observation that "Heard melodies are sweet, but those unheard/ Are sweeter," we would rather hear the world's Niagara of song, noise, and talk. Sounds thicken the sensory stew of our lives, and we depend on them to help us interpret, communicate with, and express the world around us. Outer space is silent, but on earth almost everything can make sound. Couples have favorite songs, even a few bars of which bring back sweet memories of a first meeting on the boardwalk in Atlantic City, or the steamy summer nights in a Midwestern town when, as teenagers, they sat in their Chevies at the A & W Root Beer stand, burning up hours like so many dried leaves. Mothers sing their babies to sleep with lullabies that rock and soothe, not just cradlesongs, but cradles *of* song. Music rallies people to action, as civil rights marches, Live Aid concerts, political demonstrations, Woodstock, and other mass communions have shown. Work songs and military cadence calls* make long marches or repetitive tasks less boring. Solo joggers, fast-walkers, people schussing on cross-country ski machines, astronauts pedaling stationary bikes in space, leotard-clad aerobics classes, all get psyched up from exercising to loud music that has a regular, pounding beat. A campfire wouldn't be as exciting if it were silent. And, when the campers launch their floating candles upon the lake at sunset at the end of the summer, they usually accompany the ritual with a hymn-like song of devotion to camp and one other. People want certain foods (potato chips, pretzels, cereals, and the like) to crunch; noise is an important ingredient in the marketing of such foods. Music accompanies weddings, funerals, state occasions, religious holidays, sports, even television news. Paid choirs sing poignant anthems to homeowner's insurance, laundry soap, and toilet paper. On a busy street at rush hour, despite the growl of traffic and the gyrations of thousands of hurrying strangers, we can still recognize the voice of a friend who comes up behind us and says hello. As we stroll along the reimagined streets of Williamsburg, Virginia, we hear a melodic clanging and recognize at once the sound of a blacksmith hammering on an anvil. Sitting in a chair in the living room, idly stroking the cat while sunlight streams through a window rimed with frost, may be relaxing, but when we hear the cat purr loudly we feel even more contented. Most restaurants serve obligatory music with every course; some even hire violinists or guitarists to stand at your table and ladle out enormous helpings of music as you chew. In the lobbies of hotels in India, and on the slate patios of Houston, wind chimes tinkle in the breeze. During so-called silent hours, the inmates of

---

*Carol Burke, a folklorist researching military marching chants, sent me this typical one. Most of them, she informs me, are equally crude, repetitive, and insulting.

Rich girl uses Vaseline
Poor girl uses lard
But Lulu uses axle grease
And bangs 'em twice as hard

Bang, bang Lulu
Bang away all day
Bang, bang Lulu
Who ya gonna bang today?

Rich girl uses tampons
Poor girl uses rags
But Lulu's cunt's so goddamn big
She uses burlap bags.

Bang, bang Lulu
Bang away all day, etc.

Alcatraz managed to whisper into the empty water pipe that led from sink to sink and then put an ear to the pipe to hear. Hikers llama-trekking along Point Reyes National Seashore in California, or climbing the boulder face of Mount Camelback in Pennsylvania, revel alike in the sounds of birds, river rapids, skirling wind, dry seedpods rattling on the trees like tiny gourds. In the robust festivity of a dinner party, a waiter pours a luscious Liebfraumilch, whose apricot blush we behold, whose bouquet we inhale, whose savory fruitiness we taste. Then, wishing one another well, we clink our glasses together because sound is the only sense missing from our full enjoyment of the wine.

What we call "sound" is really an onrushing, cresting, and withdrawing wave of air molecules that begins with the movement of any object, however large or small, and ripples out in all directions. First something has to move—a tractor, a cricket's wings—that shakes the air molecules all around it, then the molecules next to them begin trembling, too, and so on. Waves of sound roll like tides to our ears, where they make the eardrum vibrate; this in turn moves three colorfully named bones (the hammer, the anvil, and the stirrup), the tiniest bones in the body. Although the cavity they sit in is only about a third of an inch wide and a sixth of an inch deep, the air trapped there by blocked Eustachian tubes is what gives scuba divers and airplane passengers such grief when the air pressure changes. The three bones press fluid in the inner ear against membranes, which brush tiny hairs that trigger nearby nerve cells, which telegraph messages to the brain: We *hear*. It may not seem like a particularly complicated route, but in practice it follows an elaborate pathway that looks something like a maniacal miniature golf course, with curlicues, branches, roundabouts, relays, levers, hydraulics, and feedback loops.

Sound is transmitted in three stages. The outer ear acts as a funnel to catch and direct it, though many people lacking outer ears hear just fine (as one usually can even wearing a hat or helmet). When the sound waves hit the fanlike eardrum, it moves the first tiny bone, whose head fits in the cuplike socket on the second, which then moves the third, which presses like a piston against the soft, fluid-filled inner ear, in which there is a snail-shaped tube called the cochlea, containing hairs whose purpose is to signal the auditory nerve cells. When the fluid vibrates, the hairs move, exciting the nerve cells, and they send their information to the brain. So, the act of hearing bridges the ancient barrier between air and water, taking the sound waves, translating them into fluid waves, and then into electrical impulses. Of all the senses, hearing most resembles a contraption some ingenious plumber has put together from spare parts. Its job is partly spatial. A gently swishing field of grain that seems to surround one in an earthy whisper doesn't have the urgency of a panther growling behind and to the right. Sounds have to be located in space, identified by type, intensity, and other features. There is a geographical quality to listening.

But it all begins with quivering molecules of air, each being jostled into the next, like a crowd pressing forward into a subway. The waves they set up have a certain frequency (the number of compressions and relaxations in each second), which we hear as pitch: The greater the frequency, the higher pitched we find the sound. A large part of a sound registers as loud. Sound travels through the air at 1,100 feet per second, significantly slower than the speed of light (186,000 miles per second). That's why, during a thunderstorm, one often sees a flash of lightning and hears the thunder a few moments later. When I was a Girl Scout, we learned to start counting seconds right after we saw the lightning flash, stop when we heard the thunder, then divide by five to find out how many miles away the lightning was.

What we hear occupies quite a large range of intensities—from the sound of a ladybug landing on a caladium leaf to a launch at Cape Canaveral—but we rarely hear the internal workings of our body, the caustic churning of our stomach, the whooshing of our blood, the flexing of our joints, our eyelids' relentless opening and closing. At most, if we're wearing earplugs, or have one ear pressed against a pillow at night, we might hear our heartbeat. But for a baby in the womb the mother's heartbeat performs the ultimate cradlesong of peace and plenty; the surflike waves of her respiration lull and soothe. The womb is a snug, familiar landscape, an envelope of rhythmic warmth, and the mother's heartbeat a steady clarion of safety. Do we ever forget that sound? When babies begin talking, their first words are usually the same sound repeated: Mama, Papa, boo-boo. New parents can even buy a small box to set in the crib, which thrub-dubs a recording of a strong, regular heart rhythm at about seventy beats a minute. But if for experimental purposes the boxed heart is set faster than normal, so that it suggests an unhealthy mother, or a mother under stress, the baby will become agitated. Mother and child are united by an umbilical cord of sound.

Nothing was as perfect as that sojourn in the womb, when like little madmen we lay in our padded cells, free of want, free of time. A newborn, nursing at its mother's breast, or just being held close, hears that steady womb-beat, and life feels continuous and livable. Our own heartbeat reassures us that we are well. We dread its one day stopping, we dread the heart-silence of those we love. When we lie with our lover in bed in the morning, cuddling and dozing, pressed tight as two spoons, we feel his or her heartbeat and warmth enveloping us and are at peace. *How are you really feeling, deep in your heart?* we ask. *My heart is broken,* we answer, as if it were a block of chalk hit by a sledgehammer. Intellectually, we know that love, passion, and devotion do not lie in any one organ. A person isn't necessarily declared dead if their heart stops; brain death is the clincher. Yet when we speak of love, we use the robust metaphor of the heart, and everyone understands it. There is no need to explain. From our earliest moments, the heart measures our lives and our loves. In films, a tense, fast heartbeat is often mixed in with the musical score for scenes designed to be scary. But there are also films, like *Murmur of the Heart,* about the at-one-point-incestuous relationship between a mother and her son, where a soft, regular heartbeat enters the music to underscore the complexly loving relationship. Poems have traditionally been written in iambic pentameter, which sounds like this: ba-BUM, ba-BUM, ba-BUM, ba-BUM, ba-BUM. Of course, there are many other meters in which to write, and these days most poets don't write in formal meter at all. But there's something innately satisfying about reading a poem written in iambs. For one thing, we tend to get around in iambs; it is the rhythm of a casual stroll. But it also locks up the heartbeat in a cage of words, and we, who respond so deeply to heart sounds, read the poem with our own pulse as a silent metronome.

PHANTOMS AND DRAPES

Even those of us who damn the intrusive banalities of Muzak—consider a romantic, oceanside restaurant where you have to endure a long, sappy instrumental version of "Danny Boy" three times before paying the check sets you free—know that the brain makes its own Muzak from what it considers normal and unthreatening. Office sounds, traffic noise, heating and air-conditioning gusts, voices in a crowded room. We live in a landscape of familiar sounds. But if you're all alone at night, a familiar sound may leap out at you like a thug. Was that a screen-door hinge being opened by an ax murderer, or just a branch creaking? We hallucinate sounds more often than sights. There are auditory mirages, which vanish without trace; auditory illusions that turn out to be something other than they seemed; and, of course, voices that speak to saints, seers, and psychotics, telling them how to act and what to believe. "Listen to that little voice inside you," we say, as if the conscience were a gnome living below the sternum. But when otherwise normal people are pursued by a voice—the call of a small boy, for example, as Anthony Quinn reports hearing in his autobiography—then, like Quinn, they seek psychiatric help. Sometimes it isn't a voice, but music, people hear, hallucinating so relentlessly they think they're going mad. A doctor writing in *Australian Family Physician Magazine* in 1987 reported two cases he'd seen of severe musical epilepsy, which he thought were probably the result of a stroke affecting the temporal lobes of the brain. One of the women heard "Green Shamrock of Ireland" playing over and over in her head, and took medication to at least quiet it down some; the other, who lived until she was ninety-one and preferred the music to drugs, heard medleys of such songs as "Daisy," "Let Me Call You Sweetheart," "After the Ball," and "Nearer, My God, To Thee." The deep-dyed fright of this disorder is its hooliganism.

On the other hand, we sometimes *want* a sound to leap out at us. We want our baby's colicky cry from the other end of the house to wake us from a deep sleep, even if a louder and more abrasive sound—a garbage truck engorging, say—will not. At a busy cocktail party in a room with a low ceiling and poor acoustics, sound waves hit the wall and bounce back rather than being absorbed, and you feel as if you're in the center of a handball court in the middle of a game. Yet you can slice straight through all the noise to hear one conversation taking place between your spouse and a flirtatious stranger. It's as if we had zoom lenses on our ears. Our ability to move some sounds to the almost unnoticeable rear and drag others right up front is truly astonishing. It is possible because we actually hear things twice. The outer ear is a complicated reflector, which takes sound and hurls some of it straight into the hole; but a tiny fraction of the sound is reflected off the top, bottom, or side rims of the outer ear and directed into the hole a few seconds later. As a result, there is a special set of delays, depending on which angle the sound is coming from. The brain reads the delays and knows where to locate the sound. Blind people use their ears to map out the world by tapping with a cane and then listening carefully to the echoes. There are also times when we wish sound to preoccupy us enough to drive out conscious thought. What could be more soothing than sitting on a balcony and hearing the ocean rhythmically caressing the shore? White-noise machines fill a sleeper's room with an aerial surf, which is often just enough to free the mind from thought's clutches.

When I walked into my house last evening, I heard a noise that puzzled me at first, a sporadic creaking and almost inaudible rattling. After a few moments, I realized what it was: a field mouse writhing in a trap under the kitchen counter. Pulling back the yellow curtain, I saw him. The trap was supposed to have broken his neck fast and clean, but it had caught him across the stomach instead; without crying out or whining, he was urgently wrestling with wood and springs. Then his turmoil stopped for good. Lifting the mouse, trap and all, with a pair of fireplace tongs, I placed it carefully in a bag and put it out in the subzero garage. I'm sure he froze his fluff last night, a Scott of the Antarctic nodding as the heat-dreams fled. A homeowner needs the bloodlust of a tabby, and I don't have it. Once, at the stable, I saw a razor-boned cat harrowing a mouse until the ruin of its bloody carcass whined and thrashed, but would not quite die. The cat was following its instinct, and they were both playing out their roles in Nature, which neither gives nor expects mercy. The stable owners kept the cat specifically to hunt mice. It was not for me to intrude. But, when the cat began flaying the mouse remains, I went out back to settle my flesh-crawl by listening to the drum of ice water melting *splosh-thud* on scattered hay. Perhaps I shouldn't have been so upended by the scene of Nature, "red in tooth and claw," as Tennyson puts it. But what would I have gained by waiting out the bloody finish, the spreading wide of the ribs till they arched like open wings, the hot red jams and afterglow wiped thin across the stale cement? Instead, I focused hard on one sound— the ice water dripping onto the hay—and in a few moments relaxed enough to be able to get on with my day. I had used sound as an emotional curtain.

JAGUAR OF SWEET LAUGHTER*

We open our mouths, force air from our lungs into our larynx, our voice box, and through an opening between our vocal cords, which vibrate. And then we speak. If the cords vibrate quickly, we hear the voice as higher pitched, a tenor or soprano; if slowly, we hear an alto or bass. It seems so simple, but it's made it possible for empires to rise and fall; for children to reach small workable armistices with their parents; for corporations to control a nation as if it were a great big wind-up bathtub toy; for lovers to run the emotional rapids of courtship; for societies to express their loftiest dreams or lowest prejudices. Many of these qualities we find branded into the words themselves. Language records the fashions and feelings of a people. When William the Conquerer invaded England in 1066, he im-

---

*A creation myth found in the Popol Vuh, a book sacred to the Maya, explains that the first human creatures to appear on earth were "Jaguar of Sweet Laughter," "Black Jaguar," "Jaguar of the Night," and "Mahucutah, the Not-Brushed," with one thing in common: all could speak.

posed French customs, laws, and language, many of which we still use. The class-conscious French elite thought the subjugated Saxons uncouth and crude, and the Saxon language even at its most polite coarse and rude, first because it wasn't French, second because it was blunt. Hence, the French-derived word "perspiration" was considered polite, whereas the Saxon "sweat" was not; the French "urine" and "excrement" were polite, while the Saxon "piss" and "shit" were not. The Saxon word for lovemaking was "fuck" (from Old English *fokken*, "to beat against"),* but the French used the word "fornicate" (from the Latin *fornix*, a vaulted or arched basement room in Rome which prostitutes rented; it became a euphemism for brothel, and then a verb that meant to frequent a brothel, and finally the act performed in a brothel. *Fornix* is related to *fornax*, a "vaulted brick oven," which derives ultimately from the Latin *formus*, which meant simply warm). So "to fornicate" is to pay a visit to a small, warm subterranean room with arched ceilings. This obviously appealed more to French sensibility than the idea of "to beat against" someone, which must have seemed too animal and crude, the epitome of things Saxon.†

Sounds so captivate us that we love hearing words rhyme, we like their sounds to ricochet off of one another. Sometimes we prefer words to sound like what they mean, in the aural equivalent of a pun: *hiss, whisper, chirp, slither, babble, thump*. The word *murmur* makes us murmur just to say it, which is why these lines by Alfred, Lord Tennyson sound so perfectly full of a summer glade:

> *The moan of doves in immemorial elms,*
> *And murmuring of innumerable bees.*

The Greeks called this phenomenon "onomatopoeia," but there are forms of it so subtle that their origin has disappeared into etymological history. For example, the word "poet" comes from an Aramaic

---

*Another Saxon word for having sex was *swyve*, which the British still sometimes use.
†Ultimately documents began doubling up their terms to include both French and Saxon, and that's how legalese has stayed to this day, as in the phrases "let and hindrance," or "keep and maintain."

word that denotes the sound of water flowing over pebbles. And when we call an incompetent doctor a "quack," we're using a shortened version of the Dutch word *kwakzalver*, which literally means one who is always quacking about his salves or remedies. The way we pronounce words singles us out, gives us a sense of local or national identity, draws the rough threads of immigrant pronunciation into one reasonably smooth fabric. When people need a fresh vocabulary to deal with new challenges, terrain, or social climate, a dialect emerges. Dialects are fascinating because you can overhear in them the evolution of a familiar language, something that usually sprawls through centuries. The national language of Bermuda is English, and locals will talk to you in standard British English laced with slang gleaned from American TV, but among themselves they use a dialect not as syncopated as Jamaica-talk, but arcane and colorful all the same. "I'm gonna go ron my skirt's gates tonight and get some eez," a young Bermudian says to his friend, meaning that he's going over to his girlfriend's house to make love to her. But he needs to borrow a bike. "Can I borrow your blade?" "Don't ax about my blade, it's got a flat," his pal replies. Across the road, a pretty Bermudian girl "cuts her eyes" (looks malevolently) at them as she passes from one hotel building to another. "Bye, I'm vext!" the second young man says of his cantankerous girlfriend. "If that vedgy don't catcherself, I'm gonna slap her upside her head!"

Over the years we've tried to teach many different kinds of mammals to speak the way humans do, and though some small success has been reached with primates, dolphins, and harbor seals, we haven't had much real luck. Our ability to speak is special. We can talk for the same reason we choke so easily: Our larynx lies low in the throat. Other mammals have a voice box high in the throat, so that they can continue breathing while they eat. We can't. Remember the ventriloquist's greatest feat? Appearing to drink water and make his dummy talk at the same time. When we swallow, food slides past the trachea; if it catches there, it blocks air to the lungs. Many of us choke every year, and there's no one who doesn't know the sensation of almost choking. "It went down the wrong pipe," we gasp, perhaps lifting our arms over our head to open the airway wider. The Heimlich maneuver uses air stored in the lungs to pop the trapped food back out of the trachea. Just consider what a bad design feature this was for us. In the course of evolution, speech must have been so crucial to survival that it was worth the risk of choking.

Even if other mammals had a low larynx and a tongue in the position that would allow them to make the identical sounds we make, they would need a special part of the brain, called Broca's area, to process speech the way we do. My last answering machine had a computerized voice that gave me directions and told me what calls had arrived. I named him "Gort" after the robot in the old Michael Rennie sci-fi movie *The Day the Earth Stood Still,* because his overly flattened male voice—half zombie, half butler—sounded like an outtake from the movie. Whenever there was a power surge, Gort's logic got scrambled and he became so unreliable that I finally had to retire him. My new machine, whom I call "Gertie," speaks to me in an even flatter but female voice, which sounds uneducated and sluttish. In action, both Gort and Gertie sound subservient and unthreatening, and I suppose the manufacturers feel that's a plus. In the cockpits of large airplanes, I've heard the *annunciator*'s computerized warning—almost always a slightly sultry woman's voice*—saying such urgent things to the pilot as *"Fly up! You're too low. Fly up! You're too low,"* or reminders such as *"Your flaps are down."* The synthesized cockpit voices sound a little more lifelike because they have inflections and modulations, but computer voices in general still sound artificial. I'm sure that will change one day soon, and we'll chat amiably with articulate computers like Hal in Arthur C. Clarke's *2001.* It's only taken so long because speech is more complex than the sum of its parts. We can feed the word "top" into a computer as *t-ah-p,* but who speaks as clearly as a BBC announcer? Yet we're able to understand people talking so fast that the phonemes blur, so slowly that they drawl, in different tones, at different pitches, and with different accents. One man's *park* is another man's

---

*Research has shown that a quiet woman's voice got a pilot's attention faster than a man speaking quietly or a man or woman speaking loudly.

pahk. We make sense of one another with amazing agility, although we do occasionally have to work at it. As hard as it is for many native English speakers to understand Shakespeare's English, it's equally difficult for an American of one region to understand an American from another, since dialects are, in part, changes in the pronunciation of familiar words. Once when I was in Fayetteville, Arkansas, I asked my host if there were any spas around. I knew of the famous Hot Springs in the southern part of the state, and I thought visiting it might be a pleasant way to spend an afternoon. "Spas?" he said in a thick Arkansas accent. "You mean Russian agents?"

LOUD NOISES

One fall semester a few years ago, I accepted an appointment as a visiting professor at a college in a small leafy town in Ohio. The only visiting faculty housing was a suite in a sophomore boys' dorm, whose residents found a woman living in their midst—however discreetly—too much of a temptation. It was still brutally hot in Ohio, but almost every night someone crept up to the fusebox outside my door and threw the circuit breakers, so that my air-conditioning and all other electrical appliances loudly stopped; when I opened the door to reset the fuses, I heard scurrying and giggling down the hallway. Whenever I passed the peephole in my door, I saw an eye staring back in at me, so I covered the hole with masking tape. Twice I woke up to see a young man hanging upside down in front of my living-room window while he illegally spliced into my television cable, reducing my signal to sand. And, without fail, at nine every morning an Armageddon of heavy-metal rock began that lasted well into the night. The one sure thing I learned about sophomore boys is that they're all decibel and testosterone. Not only did their stereo music throb through the walls, it was physically painful to walk down the hallway toward the torture-level noise, and knocking on a door meant removing one hand from over an ear. The door usually opened onto a smoky room in which girls were quickly rearranging themselves and liquor or drugs hurriedly disappearing.

The diabolical noise didn't seem to bother any of them. At that volume, it was barely decipherable as music. In part, they were prematurely deaf, as frequently happens these days among loud-rock addicts. But many teenagers like to listen to music played at such high and distorting levels that it ceases to be anything but loudness. I think the loudness must excite them in an erotic way. Unfortunately, hearing can be permanently destroyed by loudness. Researchers have taken photographs of cochlear hair cells irrevocably damaged after only one exposure to a very loud noise.* Playing a ghetto-blaster at full tilt on a calm afternoon in a quiet retreat, or on the streets of a busy city, is probably more an act of aggression and dominance than of love for music: anyone within earshot will have his personal territory invaded, his peace of mind slit open.

Arlene Bronzaft, a psychologist, discovered that exposing children to chronic noise "amplifies aggression and tends to dampen healthful behavior." In a study of pupils in grades 2–6, at PS 98, a grade school in Manhattan, she showed that children assigned classrooms in the half of the building facing the elevated train tracks were eleven months behind in reading by their sixth year, compared to those on the quieter side of the building. After the N.Y. City Transit Authority installed noise abatement equipment on the tracks, a follow-up study showed no difference in the two groups. Parents don't stop to worry about which side of a building their child is going to be sitting on, and yet an eleven-month retardation in the course of only four years of school is disastrous. A child would have to struggle hard to catch up. And we wonder why kids can't read, we wonder why the drop-out rate is so high in New York. Jackhammers, riveting, and other construction noises are part of what we associate with life in big cities, but by hanging steel-mesh blankets over the construction site to absorb sound it is possible to erect a building quietly. As civilization swells, even sanctuaries in the country could

---

*Finnish researchers studying diet and heart disease discovered that a low-fat diet can improve hearing. Apparently, high cholesterol, high blood pressure, cigarette smoking, and drinking too much caffeine, which can slow up the circulation, limit blood flow through the ears, too. When rats on low-fat diets were exposed to loud noises they didn't have as much ear damage.

become too clattery to endure, and we may go to extremes to find peace and quiet: a silent park in the Antarctic, an underground dacha.

"Without the loudspeaker, we would never have conquered Germany," Hitler wrote in his *Manual of German Radio* in 1938. When we think of noise, we picture loudspeakers, radios that sound like front-line armaments, subways thundering and rattling. What is noise? Is it simply random, pain-level sound? Technically, noise is a sound that contains all frequencies; it is to sound what white is to light. But the noises that irritate us are sounds loud or spiky enough to be potentially damaging to the ear. Because a loud noise grates on our psyche, or actually hurts, we want to get away from it. But there are also nonthreatening sounds we just don't like, and we tend to classify them as noise, too. Musical dissonance, for instance. In 1899, when audiences first heard Arnold Schönberg's revolutionary "Transfigured Night," they thought it closer to organized noise than to music. *Noisy!* one passenger yells to another across the narrow aisle of a small commuter plane, like the Metroliner or Beech 1900, as the props burr, acute as a dentist's drill, and then become a denser throbbing near the bone. When someone scrapes his fingernails across a chalkboard, we twitch and convulse. So many people around the world get the willies when they hear that blackboard sound that it must not be simply a learned response, but something biological. Neurologists have suggested that it may be a relic of our evolution, when shrieks of terror alerted us to sudden doom. Or perhaps it's too much like the sound of a predator's claws skidding gently along the rock just behind us.

THE LIMITS OF HEARING, THE POWER OF SOUND

At the peak of our youth, our ears hear frequencies between sixteen and 20,000 cycles per second—almost ten octaves—beautifully, and that encompasses a vast array of sounds. Middle C is only 256 cycles per second, whereas the principle frequencies of the human voice are between 100 cycles per second for males and 150 for females. As we age and the eardrum thickens, high-frequency sounds don't pass as easily along and between the bones to the inner ear, and we start to lose both ends of the range, especially the high notes, as we may discover when we listen to our favorite music. Humans don't hear low frequencies very well at all, which is merciful; if we did, the sounds of our own bodies would be as deafening as sitting in a lawn chair next to a waterfall. But, even though we may be limited to a certain range of hearing, we're skilled extenders of our senses. A doctor listens better to a patient's heart with a stethoscope. We hang microphones in unlikely places: beneath boats to record whale songs, inside the body to record blood flow. We "hear" from the deep reaches of space and time by means of radio telescopes. Bats and bottlenose dolphins have evolved ingenious uses of sounds that are inaudible to us, and which we later invented. Doctors often rely on a form of echolocation, known as ultrasound and consisting of over 20,000 cycles per second, to help diagnose tumors. The first view a pregnant woman gets of her baby is usually an ultrasound picture. Engineers use ultrasound to test the flyability of airplane parts. Jewelers use ultrasound to clean precious gems. Sports medicine uses ultrasound to help heal sprains. And, of course, the Navy uses echolocation in submarines, though they call it sonar. You can buy a flea collar for your dog or cat that uses high-frequency sound waves to annoy fleas and ticks so that they'll vacate your pet, who supposedly doesn't hear the siren any better than you do. We may say "I'm all ears," but we tend to cock our heads or cup an ear with one hand to help out, and, when hearing fades, we aid our ears with resoundingly small electronic speakers. The original hearing aids were as large as lamp shades and only added twenty decibels; now they are small and discreet and much more powerful. But, in amplifying the world, they don't select what's meaningful from it, what needs to be heard from the pour of sheer noise.

In a cardiac intensive care unit's jungle of wires and monitors, small lights blink like the eyes of wild animals, and human hearts reveal their fury in tiny monotonous beeps. When someone's heart begins to gabble, alert technicians hear the change and come running. But researchers at Michigan State are proposing more complex and subtle monitors, ones that will produce a series of notes, not just beeps. The changing melody of each heart would offer subtle clues to its condition. Because we're used to associating the heart with sound, this doesn't strike us as particularly farfetched. However, the researchers' other proposed use of sound—to hear chemical abnormalities in a patient's urine—does, and they've borne the brunt of endless jokes about their study of musical pee.

We think of sound as something fey, lighter-than-air, an insubstantial thing, not a force with muscle. But at Intersonics, Inc., in Northbrook, Illinois, they've begun using sound to lift objects, in what they refer to as "acoustical levitation." Most objects up to now have been levitated aerodynamically or electromagnetically. Ultrasound can lift objects, too. Four acoustic transducers, emitting ultrasound waves, are arranged so that they direct narrow beams to a central spot. Where the beams intersect, an invisible stockade is created in which small objects can be suspended. Although the sound is louder than that of a jet engine, adults don't hear it. While they're floating, the objects don't feel any acoustic force, but if they drift to the side of the stockade walls, then the sound police push them back in place. Unaware of their cage unless they try to leave it, the objects seem to float in the abracadabra realm of flying carpets. But it is not a parlor game to industry, for whom this ideal crucible allows them to hold an object in place without touching or contaminating it. Ultrasound beams are powerful enough to heat a small space to the temperature of the sun, or shatter and rearrange molecules, layers of which can be stacked like flapjacks. Scientists are hoping to use ultrasound to create new glasses, including perfectly uniform glass capsules to contain hydrogen fuel in nuclear fusion reactors; brilliant alloy lenses; and fabulous electronics and superconductors. One likely application is manufacturing in outer space. "Ultrasonic levitation furnaces" went aboard the space shuttles in 1983 and 1985. New metal alloys could indeed be made of very high-temperature materials, since there would be no crucible to melt.

DEAFNESS

John Cage once emerged from a soundproof room to declare that there was no such state as silence. Even if we don't hear the outside world, we hear the rustling, throbbing, whooshing of our bodies, as well as incidental buzzings, ringings, and squeakings. Deaf people often remark on the variety of sounds they hear. Many who are legally deaf can hear gunfire, low-flying airplanes, jackhammers, motorcycles, and other loud noises. Being deaf doesn't protect them from ear distress, since humans use their ears for more than hearing. As anyone who has had an inner-ear infection knows, one of the ear's most important jobs is to keep balance and equilibrium; the internal workings of the ear are like a biological gyroscope. In the inner ear, semicircular canals (three tubes filled with fluid) tell the brain when the head moves, and how. If you were to half fill a glass with water and swirl it in a circle, the water would spin around, and, even after you stopped, the water would continue swirling for a little while. In a similar way, we feel dizzy even after we've gotten off of a merry-go-round. Not all animals hear, but they all need to know which way is up. We tend to think of the deaf as people minus ears, but they're as much prey to ear-related illnesses as hearing people are.

Despite all the folk wisdom about how important hearing is (including Epictetus the Stoic's 2,000 year-old axiom: "God gave man two ears, but only one mouth, that he might hear twice as much as he speaks"), most people, given a choice, would rather lose their hearing than their sight. But people who are both deaf and blind often lament the loss of their hearing more than anything else, perhaps none so persuasively as Helen Keller:

> I am just as deaf as I am blind. The problems of deafness are deeper and more complex, if not more important, than those of blindness. Deafness is a much worse misfortune. For it means the loss of the most vital stimulus—the sound of the voice that brings language, sets thoughts astir and keeps us in the intellectual company of man.

. . . If I could live again I should do much more than I have for the deaf. I have found deafness to be a much greater handicap than blindness.*

The literature of deafness is extraordinarily rich. Writers and thinkers from Herodotus to Guy de Maupassant have written about their own deafness or the deafness of friends and loved ones with poignancy, eloquence, and charm. The interested reader may turn to Brian Grant's anthology, *The Quiet Ear,* a fine sampler of writings on deafness that spans the centuries and many different cultures. Mark Medoff has written a powerful play called *Children of a Lesser God,* which was recently made into an equally powerful movie. My two favorite books about deafness are *Deafness: A Personal Account,* an autobiography by the poet David Wright, and *Words for a Deaf Daughter,* a classic memoir by the novelist Paul West. From Wright, we learn that his world, though it has little sound in it, "seldom *appears* silent," because his brain translates movement into a gratifying sense of sound:

> Suppose it is a calm day, absolutely still, not a twig or leaf stirring. To me it will seem quiet as a tomb though hedgerows are full of noisy but invisible birds. Then comes a breath of air, enough to unsettle a leaf; I will see and hear that movement like an exclamation. The illusory soundlessness has been interrupted. I see, as if I heard, a visionary noise of wind in a disturbance of foliage. . . . I have sometimes to make a deliberate effort to remember I am not 'hearing' anything, because there is nothing to hear. Such non-sounds include the flight and movement of birds, even fish swimming in clear water or the tank of an aquarium. I take it that the flight of most birds, at least at a distance, must be silent. . . . Yet it *appears* audible, each species creating a different "eye-music" from the nonchalant melancholy of seagulls to the staccato of flitting tits . . .

West's *Words for a Deaf Daughter* frequently appears in college syllabuses, but not, as one might imagine, only in courses for or about

---

*From a letter to Dr. J. Kerr Love, March 31, 1910, from the souvenir program commemorating Helen Keller's visit to Queensland Adult Deaf and Dumb Mission in 1948.

the deaf. Lavishly written, with much wit and phenomenological devotion, it also appeals to students of philosophy and literature as a jubilant hymn to language and life. Told in the second person throughout, it addresses and at times impersonates West's deaf daughter Mandy. And, unlike many memoirs about handicapped children, it isn't at all maudlin, but rompy, poetic, and concerned with the struggle we all wage to know ourselves and to make ourselves known. These books allow one to eavesdrop on the inner life of the deaf, a special privilege, since many people assume the deaf, especially if they don't read or write, think differently, dwelling in a no-man's-land between concept and word. But, as the literature of the deaf makes clear, ideas and emotions find their way through with surprising ingenuity, whether in English, Ameslan, or some other language, from silence to the inner world where words can be "heard."

ANIMALS

An ancient Chinese proverb says: "A bird does not sing because it has an answer—it sings because it has a song." Few animal sounds are as beautiful as bird song. Once you've heard a whippoorwill throwing the boomerang of its voice across the summer marshes, you listen with a new sense of privilege. Baby birds aren't born knowing their song; they learn the song of their parents. If you raised some birds away from their parents and whistled a different song—the opening notes of Beethoven's Ninth, say—then they would learn your song, and neighbors might well call them "the Beethoven birds." Until they get the knack of making real songs, baby birds often babble and chatter and make a lot of noise that doesn't seem to mean anything. Like human babies, they are discovering the shock of being able to make sounds at all; eventually they learn to control the sounds, and they practice. A voice is an elaborate instrument, which one can use without knowing much about it. But to make sense with it, you really need to know its limits and capabilities. Hence the babblings. Birds speak dialects, as people do. A New Hampshire crow that hasn't traveled won't respond to the call of a

Texas crow, but crows from different regions get to understand each other just as fiddlers from different states do when they meet at a convention in the Ozarks.

Some animals hear in much higher or lower ranges than we do, and with a delicacy and finesse that's astonishing. A dog can tell the difference between the sound of its master's footsteps and those of other family members or visitors. My family once had a dog that could tell the sound of my mother's car engine from any other traffic going by the house. In department stores all across America one can now buy a pair of what look like miniature foghorns, which attach to each side of a car. When the car goes about 35 mph, the wind rushing through the horns makes a high whistle that alerts deer, dogs, or other animals to get out of the way. It's too high to annoy a human ear, but to a dog napping in the road it is like an air-raid siren. Deer are nearly silent, but they hear well. An experimenter in New Zealand was recently able to cause female red deer to go into heat by playing the sound of a male red deer's mating roar. Fish don't have outer ears, but they hear vibrations through the water as we hear sounds traveling through air. Some animals can move their ears like small radar dishes, without moving their heads. I've seen deer, cats, and horses run through arpeggios of ear twitching. Thanks to a clever arrangement of their ears—one slightly higher than the other—nocturnal owls can pinpoint a sound to within one degree, and the edges of their feathers are softly fringed to muffle the sound of their approach when they are hunting. It might be more convenient to have just one centrally located ear, but having two makes it easier to locate a sound, just as having two eyes provides depth perception. African elephants have big floppy ears that mainly pick up sounds from below, and they produce a low-frequency infrasound too low for us to hear, with which they communicate.* Insects often

---

*In a letter to the editor of the *National Geographic* (December 1989), Armand E. Singer reports that "I was riding an elephant in the Terai jungle of Nepal when I heard, so low-pitched as to be almost inaudible, a vague thudding like that of a distant diesel generator. It turned out to be from my elephant, expressing fear of a nearby rhino whose scent it had caught."

have ears on unlikely parts of their bodies, such as on their legs or under their wings.

I once knew an aging cat who, when she went into heat, kept meow-screaming *"Now! Now! Now!"* over and over like a berserk harmonica player as she staggered around the apartment, occasionally stopping to thrust her rump high in that feline invitation to mating known as lordosis. Few sounds are as lovely as those made by the tree frogs in Bermuda, Puerto Rico, and other sunny isles. Often not more than an inch long, such frogs sweetly call through the night like tuneful thumb harps. It's thought that the coqui frogs of Puerto Rico locate sounds by using their lungs. Sound waves hit the sides of the frog's body, and travel to the eardrum on a pathway through the lungs. In these days of superspecialization, we assume that the body specializes, too, evolving each part for one purpose. But as it turns out, some parts have various chores. Not only frogs, but some snakes and lizards as well, hear through their lungs; in porpoises and dolphins, sound is believed to travel through an oil-filled lower jaw. Not all animals use sound just for hearing. Sperm whales, bottlenose dolphins, and others may be using sound as a weapon. It is thought that they stun their prey with loud "bangs," the blasts from which can even cause a small fish like an anchovy to hemorrhage internally.

Tonight the crickets are loud and furious, rubbing their wings into strident song. They seem to be singing in unison, but that's just an accidental felicity. I'm not hearing them talk to one another at all, since crickets communicate in the ultrasonic range, too high for human ears. What I'm hearing is accidental and to them irrelevant sounds made by their scraping wings. If I were to record the chirps and play them back for the crickets, they wouldn't answer. Animals seem to have their own lanes of sound, ones in which they communicate and to which their ears are most sensitive. If they didn't, they'd have to shriek all the time to make themselves heard above the din of other creatures.

There are auditory niches. Nature allows an animal a little decorum and privacy when it comes to its own species.* Otherwise, a warning to its brethren would also signal a predator. Of course, this doesn't always work as it should. One Central American bat, which has a special taste for the frog *Physalaemus,* stalks its prey by sound. It listens for the male frog's mating call, knowing that the louder the song is, the plumper and juicier the frog will be. This puts the frog in an appalling predicament. Full of sexual longing in the steamy tropical night, it must sing loudly to attract a mate—but if it does, it may also attract a hungry bat. And yet a poor song attracts neither.

One day in December I went with bat expert Merlin D. Tuttle to Bracken Cave in Texas, a nursery cave where millions of mother and baby bats live. Just before sunset, we sat down in the natural amphitheater of stone outside the cave and waited for the thrilling spectacle we knew was ahead of us. As a ruddy sunset began, a few bats flew out of the cave, circled to gain altitude, and flew off into the night to feed; then a few more came, and dozens after that, and hundreds after that, until suddenly the sky was thick with them. Merlin and I could feel the strong breeze they made as they identified us by echolocation and flew close to our heads without hitting us. Then Merlin swung an arm up fast and grabbed one out of the air, holding it carefully so we could look at its adaptations for echolocation, obvious even in the skin on its face: little folds and flaps that work like radar dishes.

Bats whistle or call to their prey with a steady stream of high-frequency clicks. For most of us, their vocal Braille is too high to hear, since bats click at an average of 50,000 cycles per second. In our youth, we could hear only sounds of up to 20,000. Bats click at intervals of ten or twenty times a second, and the "bat-detector" naturalists use translates the ultrasonic noises into warbles and clicks audible to human ears. Like winged megaphones, bats broadcast their voices, then listen for the sounds to bounce back at them. As they close in on their prey, echoes start coming faster or louder and,

---

*Just as there are niches in the sky, there are altitudes that various birds, bats, insects, pollen, and other fliers prefer (blue jays fly low by day when they migrate; shorebirds fly high by night), so that they won't be in extreme competition with one another.

judging the time between the echoes, a bat knows how close its prey is. The solid echoes a bat hears from a brick wall or the ground sound different from the fluid echoes of a flower or leaf. A bat can build a complete echo picture of its world, a canvas on which all the objects and animals reveal themselves in detail, down to their texture, motion, distance, and size. If you stand in a quiet yard filled with bats, the bats will be shouting very loudly; you just won't hear them. In *The Scale of Nature,* biologist John Tyler Bonner offers this way of putting echolocation into human terms:

> I can remember going through the San Juan Islands in Puget Sound in a fog. The channel between the islands is very narrow, yet it was impossible to see either shore. The ferryboat pilot first politely told all the mothers to ask their children to stop their ears. Then he blasted his horn while he leaned out the pilothouse on one side, and repeated the operation as he leaned out the other side. By judging the time it took for the echo to return, he could gauge his distance from the shore. He seemed far more composed about the process than I.

Echolocation is just one of many animal sounds beyond our hearing. Praying mantises use ultrasonics; elephants and crocodilians use infrasonics. Few animal displays are as thrilling to watch as the "water dance" of a male alligator. Stretching its enormous head out of the water, it puffs up its throat, tenses hard like a body builder, and then a rolling thunder-buster bellow splits the air, and the water sizzles all around its body, raining upward like frying diamonds. We see the water dance, but other alligators hear its infrasonic signal, made only by the males, perhaps as a courtship display or perhaps also as a full-body raspberry directed at other males. Although female alligators bellow, too, and even slap their heads on the water from time to time, they don't do a water dance. But they do read its message like seasoned code-breakers. And occasionally a male, hot and bothered and truly inspired, does a cluster of water dances—as many as eight or nine—in a long ballet of dance, song, and yearning.

We also don't hear most underwater sounds, and that leads us to assume that the vast oceans are silent, which couldn't be farther from the truth. Leonardo da Vinci once suggested dipping an oar into the water and listening, with one's ear against its handle. Fishermen in West Africa and also in the South Seas discovered the same trick. Using the oar as a kind of listening straw, you can hear the sounds of the underwater world. Some fish are a noisy lot. Sea robins, drum-fishes and many others make sounds with their swim bladders; croakers grunt loud enough to keep China Sea fishermen awake at night; Hawaiian triggerfish grind their teeth loudly; the male toadfish growls; bottlenose dolphins click and squeak like badly oiled office chairs; bowhead whales purr and twirp; humpback whales put on a songfest. The ocean looks mute, but is alive with sounds from animals, breaking waves, tidal scouring, ship traffic, and nomadic storms, locked within the atmosphere of water as our sounds are within the atmosphere of air.

How empty the world would be without animal sounds. The blackbirds quibbling like druids. Horses galloping on a soft track. The crows, which sound as if they're choking in the trees. The burbling chickadees hanging upside down from the branches. The elk's bugling, like the sound of distant war games. The metallic *ping* of nighthawks. The kindergarten band of crickets (from the Old French *criquet*, "to creak"). The electric whine of hungry female mosquitoes. The Morse code of the red-headed woodpecker.

QUICKSAND AND WHALE SONGS

Sitting on the beach in Bermuda, I decide to make quicksand in a glass. First I partially fill the glass with sand, then add water until it just covers the sand, and stir hard. The result looks solid, like firm sand, but when I stick a finger in it, it sinks fast. Quicksand is just a suspension of sand in water, sand that's become so saturated it pours like a milkshake—something temporary, not a permanent booby trap. Scary movies show people taking a wrong step, sticking deep, sinking agonizingly, and then suffocating. But that's not likely, unless you thrash about so much in panic that your body goes under and you swallow, inhale water, and drown, as you might in any swimming pool or lake. Water is denser than the human body, as is sand; and the combination makes floating doubly easy. The body is buoyant, if allowed to be. I encountered quicksand once out West, on a ranch where I was working. A cow had wandered into it and panicked trying to escape, finally drowning. When we lassoed the carcass and dragged it out, the hide was coated in a rough porridge, and the eyelids looked as if they were sewn shut with burlap. I'm sorry now I didn't wade in myself and test the waters, but at the time I listened to the cowboys' warnings. Their land-savvy never failed me, and often delighted with its intuition and clarity. They'd seen frightened horses and cattle thrash until they disappeared in the mire, and had assumed that quicksand was aggressive and always deadly.

The hypnotic crash of the waves lulls me. Bending, I press my ear against the beach and hear the waves break even sooner. The vibrations travel about ten times as fast through the ground. Were I a Kalahari Bushman, I would be sleeping on my right side tonight, ear to the ground, so I could listen for the approach of a dangerous animal; my husband would sleep on his left side, and between us there would be a small fire to keep us warm while we slept, our ears cupping the earth. Or, if I were a character in one of the old cowboy movies, I might put my ear to the tracks and listen for the sound of the oncoming mail train. Because sound waves stay inside the metal rather than dispersing into the air, I'd hear the vibrations some distance away and know the payroll, or my sweetheart, would soon be arriving.

For hours, I've been watching the ocean for signs of humpback whales, whose songs were first recorded off Bermuda by Frank Watlington, and then later by Roger Payne. When I was a graduate student at Cornell, I attended a concert that Payne gave on his cello, accompanied by whale songs that boomed, yowled, gnashed, squeaked and thrubbed, filling the large auditorium with otherworldly music, and making my bones resonate from the low-down bass notes. This wasn't the first time I'd heard whale songs; I had a record of Alan Hovhaness's musical composition "And God Created Great Whales," a piece haunted by a raga of sounds one doesn't expect to add up to song. And yet the whales do sing. In fact, they croon. Lone, inactive males start to sing during winter, the breeding season, and continue their ballads until company arrives to interrupt them. Their songs often last fifteen minutes or so, and they repeat like carols over many hours. How structured the songs are, obeying the sort of rules one associates with classical music.

What's more, the whales vary their songs. New phrases and elements arise each year, allowing the songs to evolve the way a language does. Each has half a dozen or so themes arranged in a certain order; if one theme is removed, the others still stay in their original order. When you sing "The Battle Hymn of the Republic," you may choose to leave out the verse in which soldiers have built God an altar "in the evening dews and damps," but you'll keep the rest of the verses in the right order. Within the whale songs, there are repeating phrases that follow a carefully structured whale-song grammar. Perhaps the most impressive thing about all this is that the whales not only learn the complex language, but remember it from season to season. They arrive singing the song of the previous year, like coeds returning to school in September; when new phrases and slang evolve over the season, they remember them for the following year and abandon the lingo that's out of date. They don't sing by expelling air, as one might guess. Nor do they use their blowholes in a clarinetlike way, as is sometimes shown in cartoons. Instead, they probably make their sounds by moving air around inside their heads. Like opera singers, they control their breathing very carefully, so as not to interrupt the fluency of the song. Most whales choose to do their breath-snatching in the same passages, and that allows researchers to listen for the breath spot and identify the singer.

Those who have dived among the singing whales describe the feel of the song as a drum pounding on the chest, or a pedal organ played inside the ribs. If you can't be in the water with them, you can hear and feel them singing through the wooden boards of a boat. And not only humpbacks sing. White beluga whales have such a sweet, trilling voice that early whalers called them "sea canaries." Now that their numbers are drastically reduced by pollution, the belugas are becoming the canaries in a liquid mine, warning us about the health

of the oceans. Superstitious sailors used to hear the mournful songs
of whales echoing up through the hulls of their ships, and were
enraptured. Singing whales once inhabited the Mediterranean, and
probably are the Sirens Greek myth says lured sailors to their doom
on the rocks. Coming through the wood of a boat, their songs would
be diffused in such a way that a sailor couldn't localize them; the
sounds would seem to be enveloping the ship in an eerie veil of song.
Because whales ululate in sounds unique and varied, it's a little
difficult to describe their voices, but I once wrote the following
sound poem after hearing a whale concert, and it may give a better
sense of their songs:

WHALE SONGS

Speaking in storm language,
a humpback, before it blows,
lows a mournful ballad
in the salad-krill sea, murmurs
deep dirges; like a demiurge,
it booms from Erb to Santa Cruz,
bog low, its foghorn a thick liqueur.

Crepe black as a funeral procession,
the pod glides, mummer-deft,
through galloping brine,
each whale singing the same
runaway, roundelay tune:

Dry fingers rub, drag, drub
a taut balloon. Glottal stops. Pops.
Dry fingers resume, then, ringing
skeletal chimes, they ping
and rhyme—villanelles, canticles,
even a Gregorian done on ton tongues

as, trapped below the consciousness
of air, hungry, or wooing,
or lamenting slaughter,
jazzy or appalled,
they beat against the wailing wall
of water, voices all
in the marzipany murk they swim,
invisible but for their songs.

And often they raise high
as angels' eyes a refrain
swoony as the sea, question-mad,
sad, all interrogatives, as if
trying to fathom the fathomless
reach from ladle-shaped ocean,
scurrilous surf, to breach-birth
upon beach and blue algae's cradle.

Sleek black troubadours
playing their own pipes, each body
a mouth organ, each shape a daguerreotype
of an oblate friar caroling,
they migrate, glad to chain rattle
and banshee moan, roaming the seas
like uneasy spirits, a song on their bones.

THE VIOLIN REMEMBERS

Music, the perfume of hearing, probably began as a religious act, to arouse groups of people. Drums set the heart sprinting in no time, and a trumpet can transport one on chariots of sound. As far back as we can see, people made music. The first instruments used in western music were probably just sticks or rocks thwacked together to make a beat. There would have been many occasions for them: religious dances and other rituals; to accompany work songs; as a musical way to teach lessons to the young. Mesopotamian instruments have been found dating back some 5,500 years (pipes, triangles, stringed instruments, and drums), and the Mesopotamians even devised a method of musical notation. People probably made music even earlier than that, by blowing on blades of grass held between their thumbs or banging sticks and stones together—instruments we wouldn't now be able to recognize. The Mayans played an array of intricately carved clay whistles, flutes, recorders, and ocarinas. Whistles shaped like men produced lower notes than those shaped like women. Some of them had secret chambers and could play as many as seventeen notes, others were meant to hold water while you played them, which affected the sound, and some multi-headed flutes played several notes simultaneously. According to Chinese texts, Oriental music began around 2700 B.C., when Huang Ti, the emperor, ordered bamboo pipes of the right length to be cut so that he could imitate the song of the phoenix. If one contrasts 2,400-year-old Chinese bells with a present-day Chinese flute, one finds that the tones are very similar, and nearly match on an oscilloscope. From the outset, our brains and nervous systems have led us to prefer certain intervals between sounds. Our instruments have evolved from a deep inner delight in music, but one that has boundaries. Much of what we hear strikes us as dissonance or as noise, and what falls within a certain range we find sweet, intellectually satisfying, and mellifluous.

I first learned to play the violin in junior high, and though I practiced haphazardly for eight years, I never got past the mechanical bowing, palsied vibrato, and lusterless finger work of an amateur. I loved the gritty yet oily shine of the resin, which allowed the bow to tug gently, as if dragged over a raspy cat's tongue. The strings I bought were referred to as "catgut," but of course they weren't really from a cat; the slang term dated back to an early period in violin playing, when audiences thought the strings screeched like a disemboweled tabby. "Better go buy some more catgut!" they used to jeer, and the expression caught on. Even when I was a "tweenager" (as thirteen- and fourteen-year-olds were then called), endlessly rehearsing "The Entrance March of the Peers," "The Young Prince and the Young Princess," and "Say It with Music" for school assembly, I'd heard rumors of a dark, nearly mythical violin that could virtually play itself, a violin that smoldered with caged emotion even when lying in its case. The name of it floated in my mouth like magical smoke: Strad-i-var-i-us. How often I lusted after a Stradivarius that would transmute my sandpapery sounds to pure gold. In time, I rose through the ranks to the orchestra's honored position of "first violin," which meant that I got to play the melody, which is why I chose to learn violin in the first place. I pitied the tuba players oompahing their way into oblivion. Some of them, though boys, weren't athletically built, and when they stood up they half disappeared into the shiny, heavy, hallucinating brass, as if swallowed whole by a mirrored nautilus. The percussionists made such a nerve-jangling racket, I thought they should be given a polite burial in their own kettle-drums. Nothing about the finicky, birdlike oboe appealed to me. The girls who played flute always had runny noses and looked as if they were trying to blow out a small flame when they played. The clarinets sounded too mouselike. And the idea of playing cello, viola, bass, or any of the other to-my-mind subservient instruments left me cold. I wanted to make music, and music to me was melody, a soulfully singing violin. Although I had never heard a Stradivarius up close, I heard them on records and on television, and I wondered along with everyone else what magical resin or lacquer had gone into their manufacture to produce their uniquely sultry richness. The most precious instruments in the world are still the violins made by Stradivarius. At last scientists are beginning to understand why.

Over the years, researchers have attributed the unique sound to animal fluids, special resins, a water fungus, and many other arcane potions. A more likely explanation was proposed recently by Peter Edwards and a team of researchers at Cambridge University. Using EDAX (energy dispersive X-ray spectroscopy), they showered a fragment of a cello with high-energy electrons, which allowed them to analyze the wood's ingredients. To their surprise, they found a thin layer of pozzolana—a volcanic ash from Cremona, Italy, where Stradivarius lived. The ash lay between the varnish and the wood, and Stradivarius probably applied it as a simple strengthening agent for his instruments; since it was a commonly used cement, it probably never occurred to him it could affect their tone. Of course, pozzolana alone won't produce a Stradivarius, whose age, architecture, and craftsmanship contribute to its sound. Many violinists and violinmakers insist that violins grow into their beautiful throaty sounds, and that a violin played exquisitely for a long time eventually contains the exquisite sounds within itself. Somehow the wood keeps track of the robust lyrical flights. In down-to-earth terms: Certain vibrations made over and over for years, along with all the normal processes of aging, could make microscopic changes in the wood; we perceive those cellular changes as enriched tone. In poetic terms: The wood remembers. Thus, part of a master violinist's duties is to educate a violin for future generations.

MUSIC AND EMOTION

One of the most soothing things in the world is to put your tongue to the roof of your mouth right behind the teeth and sing *la, la, la, la, la, la, la.* When we sing, not only do our vocal cords vibrate, but so do some of our bones. Hum with your mouth closed, and the sound travels to your inner ear directly through the skull, not bothering with the eardrum. Chant "om," or any other mantra, in a solid, prolonged tone, and you will feel the bones in your head, as well as the cartilage in your sternum, vibrate. It's like a massage from the inside, very soothing. Another reason it may be so conducive to meditation is that it creates an inner white noise, which cancels out extraneous noises, making your body a soundproof booth. Hebrew davening, in which the faithful bend and chant, bend and chant, has a similar effect. The drumbeat in a macumba ceremony seizes one in a crescendo of fury that climbs higher and higher, as if scaling the Himalaya of one's belief. All these sounds repeat hypnotically. Every religion has its own liturgy, which is important not just in its teachings but also because it forces the initiate to utter the same sounds over and over until they are ingrained in memory, until they become a kind of aural landscape. We are a species capable of adding things, ideas, and creative artifacts to the world, even sounds, and when we do, they become as real a fact as a forest.

The odd thing about music is that we understand and respond to it without actually having to learn it. Each word in a verbal phrase tells something all by itself; it has a history and nuances. But musical tones mean something only in relation to one another, when they're teamed up. You needn't understand the tones to be moved. Say the words "It's a gift to be simple. It's a gift to be free. It's a gift to come down where we ought to be," and nothing much happens. You might even disagree with its minimalist doctrine. Yet if you add the tuneful Shaker music that goes with it (which Aaron Copland adapted so beautifully in *Appalachian Spring**), its haunting mel-

---

*He wrote this music for Martha Graham while living in a Hollywood block house with no windows.

ody, full of enough ebullience, joy, and conviction to inspire a whole village to put up a neighbor's barn in one afternoon, will truly captivate you. When I was in Florida, at an artist's colony on a tidal estuary, one of my writing students, also a professional whistler, regaled us one evening with a whistle concert, including this Shaker tune, "Simple Gifts," and for the next week you could hear people humming, whistling, or singing its gaily hammering rhythm. *Catchy* is the right word for such a melody; it hooks onto your subconscious and won't let go. Many hymns would thrill us even if they didn't have words, but, with words, they're a double score: emotional music tied to emotional messages. It works particularly well if the hymn has a dying fall in it, a musical swoon. In Blake's "Jerusalem," that swoon comes in the third stanza, in the second syllable of the word "desire," which you have to sing as a sigh to a lower note:

> *Bring me my bow of burning gold!*
> *Bring me my arrows of de-sire!*

Few desires sound as smoldery and secular as that one, especially if you're reminded of Cupid's arrow and the double meaning of a word like "quiver." In the Christmas hymn "O Holy Night," the swoon comes right after the word "fall," in the line "Fall on your knees," and just singing it enacts the supplication. Most often hymns soar steadily in slow sweeping steps, from lower to higher notes, as the singer climbs a mystical staircase onto progressively higher planes of feeling. "Amazing Grace" is a good example of that lighter-than-air sort of hymn, full of musical striving and stretching, as if one's spirit itself were being elongated. Think lofty thoughts and sing that elevating tune, and soon enough you will feel uplifted (even despite having to sing such unmelodious words as "wretch"). Hypnotists use a similar technique when they put people into a deeply suggestive meditative trance: They often count from one to ten a few times over, telling patients to imagine themselves climbing deeper and deeper down with each number.

Like pure emotions, music surges and sighs, rampages or grows quiet, and, in that sense, it behaves so much like our emotions that it seems often to symbolize them, to mirror them, to communicate them to others, and thus frees us from the elaborate nuisance and inaccuracy of words. A musical passage can make us cry, or send our blood pressure soaring. Asked to define the feeling, we say something vague: *It made me sad.* Or: *It thrilled me.* In *Great Pianists Speak for Themselves, Vol. II,* Paul Badura-Skoda says of Mozart's Fantasy in C minor:

> What about the *emotional* content? What does the work *say* to you and me? Surprisingly, when I ask such questions in my master classes, I get rather tepid answers such as, "It is a serious work," or none at all. Then I am forced to exclaim, "Don't you realize, my dear fellows, that music is a *language* which *communicates* experience? And what experience! Life and death are involved in this *Fantasy.* May I tell you my personal interpretation of this work? The opening phrase is a death symbol: *The hour has struck—there is no escape!* The rest of the Fantasy is shock and anxiety, pages one and two, giving way then to a series of recollections: happy, serene ones, like the Adagio in D and the Andantino in B-flat major, or violent ones, full of anguish, like the two fast, modulating sections, until finally the original call returns. The inexorable fate seems to be now accepted, were it not for the heroic gesture of defiance at the very end.

Not all composers care for listeners to find such a clear program in their work, but people get so frustrated by the abstractions of music they try to elicit from it landscapes of emotions and events.

We find a profound sense of wholeness in the large, open structure of a classical composition, but it is a unity filled with tumult, with small comings and goings, with obstructed quests, with bouts of yearning and uncertainty, with insurpassable mountains, with interrupted passion, with knots that must be teased apart, with great washes of sentimentality, with idle ruminations, with strident blows to recover from, with love one hopes to consummate, with abruptness, disorder, but, ultimately, with reconciliation. One can re-create the emotional turmoil of an affair, a disappointment, a religious ecstasy, in as small a space as a concerto. *Show, don't tell!* writing teachers counsel their students. Say what one will, words rarely capture the immediate emotional assault of a piece of poignant music, which allows the composer to say not "It felt something like this," but rather "Here is the unnamable emotion I felt, and even my obsession with structure, proportion, and time, *inside of you.*" Or, as T. S. Eliot puts it in "The Dry Salvages," here is:

> *music heard so deeply*
> *That it is not heard at all, but you are the music*
> *While the music lasts.*

There are still many questions to be answered about music and emotion. In his fascinating book on music theory, *The Language of Music*, Deryck Cooke, for example, offers a musical vocabulary, spelling out the emotional effects a composer knows he can create with certain sounds. But why is this so? Do we tend to respond to a minor seventh with "mournfulness" and to a major seventh with "violent longing" and to a minor second with "spiritless anguish" because we've formed the habit of responding to those sounds in that way, or is it something more intrinsic in our makeup? Listen to Wagner's *Tristan und Isolde,* and you'll hear pent-up, soaring, frustrated emotion of an intensity that may drive you to distraction. *Yearning* overflows the music like the meniscus on a too-full glass of wine, and this is how Wagner himself described the work:

> . . . a tale of endless yearning, longing, the bliss and wretchedness of love; world, power, fame, honour, chivalry, loyalty, and friendship all blown away like an insubstantial dream; one thing alone left living—longing, longing, unquenchable, a yearning, a hunger, a anguishing forever renewing itself; one sole redemption—death, surcease, a sleep without awakening.

Another question we might ask, along with Cooke, is: If we transform music into emotion, "how closely does this emotion . . . resemble the original emotion of Beethoven? . . . There can only be one answer to this . . . about as closely as the emotions of one human being can ever resemble those of another." And, because we're not Beethoven, we hear his joyous "Gloria" in the *Missa*

Solemnis and feel joy, but probably not as passionately as he did when he wrote it. I suppose part of what's fascinating about creativity in any field is the author's necessity to share it with—or impose it on—the world. When he wrote the "Gloria," Beethoven underwent a volcanic, shriek-to-the-heavens joy, but instead of dancing around in delight, he "felt the need to convert it into a permanent, stored-up, transportable, and reproducible form of energy," as Cooke describes it, "a musical shout for joy, as it were, that all the world might hear, and still hear over and over again after he was dead and gone." The notes he jotted down "only ever were and only ever will be a command from Beethoven to blow his eternal shout for joy, together with a set of instructions . . . exactly how to do so." When we proclaim that artists live on in their work, we're usually referring to the emotional steppingstones that lead through their lives, their disembodied moods and obsessions, but most of all their senses. Beethoven may be dead, but his sense of life at that moment lives in his score at this moment, at any moment.

IS MUSIC A LANGUAGE?

Music speaks to us so powerfully that many musicians and theorists think it may be an actual language, one that developed about the same time as speech. One Harvard psychologist believes strongly that music is a kind of intelligence, an aptitude like that for words or numbers, with which we're simply born. By experimenting with brain-damaged musicians, he's been able to locate musical ability in the right frontal region of the brain. In a related experiment, researchers at the UCLA School of Medicine gave volunteers a Sherlock Holmes story to read, then music to listen to, and recorded brain activity with a PET scan. Reading excited the left hemisphere of the brain, music the right. But knowing where our passion for music lies doesn't explain how it got there. No matter how far back in history we look, we find human beings making and listening to music, but how and why did our passion for it begin? Why do we feel driven to make music? Why does music differ so much between cultures? Why do many people feel the need to live in cocoons of organized sound, to keep music close at hand? Why do we respond to music's array of abstract sounds with intense, sometimes violently felt emotions? If music evolved along with spoken language, *why* did it evolve? What was its survival value? Music is meaningful, as anyone listening to a soulful symphony or an opera by Wagner would readily admit, but what is its meaning? How do we assign a particular meaning to a piece of music? Why does music make sense even to people who don't play instruments themselves, and even claim to be tone-deaf, people who aren't particularly "musical"? Most of all: How do we understand the language of music without *learning* it? For the moment, the reasonable answer to that last question is that, like the ability to smile or analyze, it's deeply hereditary. At some point in our past it was important enough that all human beings born, no matter whether Bengalese, Inuit, or Quechua, no matter whether blind, left-handed, or freckled, were not merely *capable* of making music; they *required* music to add meaning to their lives. The newest infant responds to music, and by the time a child can toddle it can already sing songs, and even make them up. To a certain extent, music is also learned. Children in China learn to like music with small intervals and subtly changing pitches; children in Jamaica learn to like syncopated ballads; and children in Africa learn to like music with fast, intricate rhythms. One's musical preferences can be willful. Generations tend to define themselves by a music that differs from that of their parents, who usually describe the new music as noise, obscene, a waste of time, and lacking in any art. When the waltz first came in, it was thought avant-garde and scandalous.* After all, it caused men and women to hold on to each other and move rapidly, clinging wildly while their hair flew, their petticoats fluttered, and their hips rocked in unison. The same was true of swing music, which the older generations of the time found barbaric, repetitious, or just silly. What were they to make of lyrics like: "It must be jelly, 'cause jam don't shake like that"? And the tango had its own sneaky, insinuating rhythm and a sexy dance step in which a woman wraps her leg around a man's leg as if he were

---

*Lord Byron wrote a famous poem about the waltz, whose excesses he admired.

a tree and she a vanilla orchid's climbing vine. The words that accompanied all this carnal mayhem were usually sensuous, violent, and extravagantly heartrending. Here are the lyrics to a typical Argentinean tango, taken from Philip Hamburger's *Curious World:*

> All my life, I have been a good friend to everyone. I have given away everything I own and now I am all alone, ill, in my dirty and gloomy small room in my neighborhood slum, coughing blood. No one comes to see me now except my dear mother. Ah, now I realize my cruelty to her. I am at the point of death and I recognize my love for her. She is the only one who really cares for me.

In recent times, science fiction has proposed music as the Esperanto of the universe, a language which even far-flung creatures might share. *Close Encounters of the Third Kind* is perhaps the best example of a sci-fi story based on that premise. A single chord is a calling card and, at that, a mighty simple chord, based on universally shared mathematics. This is an old idea, going back to the Greeks and the music of the spheres. There has always been a connection between music and mathematics, which is why scientists have often been inordinately fond of music, especially of composers such as Bach. The composer Borodin was first and mainly a scientist, who discovered a method for combining fluorine and carbon atoms to produce new compounds. We're indebted to his inspiration for Teflon, Freon, and a variety of aerosols. His hobby was composing music. At the Fermi National Accelerator Laboratory in Illinois, there is a concert hall among the offices and labs. Some West German physicists are studying the relationship between musical composition and the mathematics of fractals. Why is music mathematical? Because, as Pythagoras of Samos discovered in the fifth century B.C., notes can be precisely measured along a vibrating string, and the intervals between notes expressed as ratios. Of course, people sang what pleased them; they didn't decide to sing in ratios. This revelation, that mathematics was secretly determining the beauty of music, must have seemed just one more indisputable proof to the mathematically minded Greeks that the universe was an orderly, logical, knowable structure. The Greeks used to play or sing their scales downward, from high to low. We prefer to sing or play ours upward, from low to high. This change really began with Christianity and the Gregorian chant, and I think it came about as a result of religious uplift and a desire for transcendence. Science fiction argues that if music is mathematical then it must be universal. For interstellar space, don't bother with verbal messages; send a fugue. To be safe, send both. When Voyager I was launched in 1977, it carried assorted messages for other planetarians to find, including a record that contains miscellaneous sounds of Earth as well as Earth's music, and instructions as to how to play the record.

Does music, then, have a grammar, like language, or its own set of mathematical laws? If it's principally mathematical, how come mathematically illiterate people still revel in it? In an essay in *New Literary History* in 1971, composer George Rochberg argued that "music is a secondary 'language' system whose logic is closely related to the primary alpha logic of the central nervous system itself, i.e., of the human body. If I am right, then it follows that the perception of music is simply the process reversed, i.e., we listen with our bodies, with our nervous systems and their primary parallel/serial memory functions." *We listen with our bodies.* Indeed, it's hard to keep our bodies still when we hear music—our feet begin tapping, our hands begin swaying, we pick up an invisible baton, or gyrate in some sketchy dance movements. In Peter Schaffer's play about Mozart, *Amadeus,* Salieri, the established and rival composer, says:

> It started simply enough: just a pulse in the lowest registers—
> bassoons and basset horns—like a rusty squeezebox. . . . And then
> suddenly, high above it, sounded a single note on the oboe. It hung
> there unwavering, piercing me through, till breath could hold no
> longer, and a clarinet withdrew it out of me, and sweetened it to
> a phrase of such delight it had me trembling.

A musical note is just pulsating air stimulating the organs in our ears. It may have various qualities, like volume, pitch, or duration, but it is still just pulsating air. That's why the deaf often enjoy music, which they perceive as attractive vibration. Helen Keller "heard" Caruso sing by pressing her fingers to his lips and throat, and she writes beautifully about holding a radio and listening to a symphony concert, responding to the different instruments as they joined in. An oscilloscope can make the tones visible. Since it displays vibration, it can reveal the acoustical properties of the tone, but there is no way it can judge the musical experience. When Duke Ellington plays piano, I hear many of the pastel, water-ice phrases of Ravel, but how could I begin to describe an Ellington piece? If you haven't heard a tone before, there's no word that will reproduce it or faithfully conjure it up. Teddy Wilson, who played piano with the Duke's band for a while, remembers how Ellington used to play the dance rhythm with his left hand while with his right he created a splash of excitement, which he describes picturesquely as "like throwing colored sand up into the air."

Countries speak their own unique languages, but whole civilizations enjoy certain forms of music, which we, perhaps too chauvinistically, refer to as western music, Oriental music, African music, Islamic music, and so on. What we mean is that each civilization seems to prefer hearing tones arranged in certain patterns according to slightly different laws. For the past 2,500 years or so, Western music has been obsessed with one polyphonic arrangement of tones, but there are many other arrangements, each as profoundly meaningful as the next and yet incomprehensible to outsiders. "The barriers between music and music are far more impassable than language barriers," Victor Zuckerkandl writes in *The Sense of Music*. "We can translate from any language into any other language; yet the mere idea of translating, say, Chinese music into the Western tonal idiom is obvious nonsense." Why is that so? According to the composer Felix Mendelssohn, it's not because music is too vague, as one might think, but rather too precise to translate into other tonal idioms, let alone into words. Words are arbitrary. There's no direct link between them and the emotions they represent. Instead, they lasso an idea or emotion and drag it into view for a moment. We need words to corral how we feel and think; they allow us to reveal our inner lives to one another, as well as to exchange goods and services. But music is a controlled outcry from the quarry of emotions all humans share. Though most foreign words must be translated to be understood, we instinctively understand whimpering, crying, shrieking, joy, cooing, sighing, and the rest of our caravan of cries and calls. I believe that, in time, they led to two forms of organized sound—words (rational sounds for objects, emotions, and ideas) and music (nonrational sounds for feelings). As Cooke observes, "both awaken in the hearer an emotional response; the difference is that a word awakens both an emotional response and a comprehension of its meaning, whereas a note, having no meaning, awakens only an emotional response." What sort of response can a few notes of music awaken? Awe, rage, wonder, restlessness, defeat, stoicism, love, patriotism.... "What passion cannot Music raise and quell?" John Dryden asks in his "A Song for St. Cecelia's Day," and then goes on to say:

> The soft complaining flute,
> In dying notes, discovers
> The woes of hopeless lovers,
> Whose dirge is whisper'd by the warbling lute.
>
> Sharp violins proclaim
> Their jealous pangs and desperation,
> Fury, frantic indignation,
> Depth of pains, and height of passion,
> For the fair, disdainful dame.

In a letter to his father, written in Vienna on September 26, 1781, Mozart said of his *Abduction from the Seraglio:*

> Now, as for Belmonte's aria in A major—"O wie angstlich, O wie feurig"—do you know how it is expressed?—even the throbbing of his loving heart is indicated—the two violins in octaves.... One sees the trembling—the wavering—one sees how his swelling breast heaves—this is expressed by a *crescendo*—one hears the whispering and the sighing—which is expressed by the first violins, muted with a flute in unison.

For Mozart, music was not only a passionately intense intellectual medium, it was one through which he felt, indeed conducted, precise emotions. The theme of the first movement of Mahler's Ninth

Symphony mimics his cardiac arrhythmia, and therefore laments his mortality. He died soon after, in the middle of writing his Tenth Symphony.

Of course, there is an odd sense in which music can't really be heard at all. Much of musical composition is tonal problem solving on a very complex scale, an effort undertaken entirely in the mind of the composer. Not only is the orchestra not necessary for that creative feat of legerdemain, it most likely will produce an inferior version of the music the composer imagines. How could Beethoven write the Ninth Symphony so brilliantly when he was deaf, people wonder. The answer is that it wasn't necessary for Beethoven to "hear" the music. Not as sound, anyway. He heard it flawlessly and much more intimately in his mind. Everyone touched by a piece of music hears it differently. The composer hears it perfectly in the resonant chambers of his imagination. The general audience hears it emotionally, without understanding its craft. Other composers hear it with an insider's knowledge of form, structure, history, and incunabula. The members of an orchestra—arranged according to instrument—hear it boomingly, from "inside," but not as a balanced work.

Some animals and people speak in music alone. For example, on the island of Gomera in the Canaries, descendants of an aboriginal people called the Guanches, about whom little is known except that they lived in caves and mummified their dead, use an ancient whistling language to communicate across the sprawling valleys. They trill and warble a little like quails and other birds, but more elaborately, and, from as far away as nine miles, they hear one another and converse as their ancestors did. *Silbo Gomero* the idiom is called, and some islanders mix it with Spanish vocabulary to make a creole of whistle and word. They find this hybrid language precise enough.

In Australia, the aboriginals have divided up their land according to a maze of invisible roads, or Songlines, across which they travel to conduct the normal affairs of their lives. Closest perhaps to the way in which bird song maps out a territory, the Songlines are ancient and magical, but they are also precise map references. The continent is crisscrossed by a labyrinth of Songlines, and the aboriginals can sing their way along them. As Bruce Chatwin describes the process in *The Songlines:*

> Regardless of the words, it seems the melodic contour of the song describes the nature of the land over which the song passes. So, if the Lizard Man were dragging his heels across the salt-pans of Lake Eyre, you could expect a succession of long flats, like Chopin's "Funeral March." If he were skipping up and down the MacDonnell escarpments, you'd have a series of arpeggios and glissandos, like Liszt's "Hungarian Rhapsodies".
>
> Certain phrases, certain combinations of musical notes, are thought to describe the action of the Ancestor's *feet.* . . . An expert songman, by listening to their order of succession, would count how many times his hero crossed a river, or scaled a ridge—and be able to calculate where, and how far along, a Songline he was.

When words and music meet in poetry or in song, each enhances the effect of the other. As our emotions flare, our speech naturally becomes more lyrical. "All passionate language does of itself become musical," Thomas Carlyle observes, "the speech of a man even in zealous anger becomes a chant, a song." This is never more evident than in the sermons of fundamentalist preachers, or the rhetoric of strident political activists, or the stanzas of Russian poets, who sing their verse. Virtually all movies these days have soundtracks and background music. The assumption must be that we're not competent to hear the world, and that we need music to supply us with quick, relevant emotions. Is this because we don't think the world is worth listening to? Is it because filmmakers wish to combine words and music for the most intense emotional effect? Or is it just that they think we're too lazy, or too shallow, or too numb to have an emotional response to what we're viewing?

MEASURE FOR MEASURE

Some facets of our biology are ideally shaped for music, which pours through them as beautifully as light through a stained-glass window. William Congreve was right: "Music hath charms to soothe a savage breast." Over the years, many people have slurred that aphorism to read *beast*, not *breast*, but Congreve didn't mean that lions are tamed by music, or cobras hypnotized by the snake charmer's flute (anyway, it's the movement of the flute and the charmer himself, not its sound, which fascinates the snake; snakes are deaf). He meant that music can calm the hearts of the most bloodthirsty of us, even against our will. Most often, our emotions are private things. We bottle them up like so many jars of peach preserves that we store on a top shelf in a hidden pantry; then, in a crisis, we reach for them, often taking off the lids on our emotions through song. People who sing and wail at wakes know how therapeutic this can be. We often vent great passion by breaking into song. Strangers who seem to share nothing, not even the same culture, can sing with a mournfulness or jubilation all understand. Manfried Klein, an Australian physiological psychologist, conducted studies in which he played passages of Bach and then measured the hand-muscle responses of a group of volunteers. Regardless of their cultural background (Japanese and American businessmen, Australian aboriginals, and others), all responded to the same passages of Bach in the same way. Next he measured hand-muscle responses when they felt joy, anger, and other strong emotions. The graphs plotted for the emotional states corresponded to those for the passages of Bach. Music seems to produce specific emotional states that all people share, and as a result, it allows us to communicate our most intimate emotions without having to talk about or define them in a loose net of words.

Our pupils dilate and our endorphin level rises when we sing; music engages the whole body, as well as the brain, and there is a healing quality to it. In World War II, it was discovered that even comatose patients could respond to music. Doctors and nurses use music to help them reach handicapped children, especially children with multiple handicaps. Autistic or learning-disabled children, who find speaking an insurmountable hurdle, frequently have less trouble communicating first in song, then transferring their facility to speech. Because music can be so uplifting and recharging, it encourages sedentary people to exercise longer and more often. The usual choice is jazz, swing, pop, or rock, whose rhythms jar our natural heart rhythm and make our blood pressure rise; we feel revved up. Music can also calm. Some therapists specialize in a course called "Guided Imagery in Music," working with blindfolded patients who are led into a relaxed state where fruitful images may form. In some cardiac intensive-care wards, angina patients listen to classical music as part of their recovery process. It both relaxes them and draws a musical blind down over the frightening scenes around them. Some doctors prescribe music for cancer patients, the elderly, the emotionally disturbed or mentally ill. And there's a large international organization of music therapists, whose most recent annual conference included sessions on "The Use of Music in Teaching Reading to Hearing-Impaired Children," "The Aging Nervous System: Problems for Music Therapists in Geropsychiatry," "Promoting Psychosocial Adjustment in Pediatric Burns through Music Therapy," "Music Therapy in the Rehabilitation of Traumatically Brain-Injured Persons," and many other intriguing-sounding topics.

To understand why music pleases us, we must ask why we feel pleasure at all. What we perceive as "pleasure" may be just the thrill of shooting the rapids on our body's "river of reward," as chemist James Olds nicknamed it. It was Olds who, when he was conducting experiments with rats, first located the brain's pleasure center. Like the rest of the body, the river of reward is a strange alloy of electricity and chemicals, and there are various ways to trigger or quiet it artificially, using electrodes or drugs. From the outset, we've evolved through a thick tapestry of rewards, so it shouldn't surprise us that quiz shows, contests, medals, and award-donating programs of every conceivable kind dominate our culture, or that addictions are so hard to break. Reward, one of the central players in the brain, wears many masks. Like a melody, it can appear in a higher or lower key, at a faster or slower pace, on a wide array of instruments; it can be simple or elaborate, and still be recognizable.

In the Addiction Research Laboratory at Stanford University, a woman sits in a soundproof room and listens to her favorite music through headphones. This happens to be a concerto by Rachmaninoff, which builds to one orgasmic crescendo after another, but other student volunteers will choose other classics, pop songs, or jazz. The choice is irrelevant as long as it sends shivers of delight through the listener. Tingles usually start at the back of the neck, creep over the face and across the scalp, dart along the shoulders, trickle down the arms, and then finally shiver up the spine. Isn't it odd that intense emotion or esthetic beauty gives us chills? When this happens, the woman in the soundproof room signals with one hand. Because she feels thrilled quite often while listening to music, she's put into a second group and tested again. This time, she's given naloxone, a drug that blocks endorphins, our natural opiates. Others being tested are given placebos. Van Cliburn begins his lusty performance of Rachmaninoff's Second Piano Concerto, then sweeps into the tight, mounting rhythms of the first crescendo, which has always made her tingle. This time the music just lies flat in her mind. Her body feels nothing. The rapture is gone.

CATHEDRALS IN SOUND

For a long time, western music was homophonic, or "same voiced," which doesn't mean that only one person sang at a time but rather that there was one melody line or voice, and the rest of the music was harmony supporting it. Usually the main melody was the highest pitched, and identified the piece. Plainsong, the religious music of the fourth century, required no musical accompaniment at all; one voice sang the simple melody to Latin words. In the sixth century, Pope Gregory I decided to govern music making; as a result, the Gregorian chant evolved, which was sung in unison. In the Middle Ages, people made the extraordinary discovery that many tones could be made at once without canceling one another out or resulting in mere noise, and polyphony was born. It seems impossible that it could have taken so long to reach that now-obvious conclusion. But music is not like vision. If you mix blue and yellow together, you lose the individual colors and make a new one; tones, on the other hand, may be combined without losing their individuality. What you end up with is a chord, something new, which has its own sound but in which the individual tones are also distinct and identifiable. It's not a blending or, as one might expect when one hears a number of people talking at once, just noise, but something of a different order. A chord "is something like an idea," philosopher of music Victor Zuckerkandl writes, "an idea to be heard, an idea for the ear, an audible idea." For colors to stay separate without blending, they have to occupy space next to one another. They can't occupy the same space. But notes *can* occupy the same space and remain separate. As Zuckerkandl reminds us, polyphony "coincided with the building of the great Gothic cathedrals, and the birth of harmony with the culmination of the Renaissance and the beginning of modern science and mathematics: that is, the two great changes in our understanding of *space*."* This may seem an odd observation, given the fact that vision is a spatial art, and music a temporal one, which "unfolds in time," a dynamic art that uses many devices, including *syncopation*, in which notes appear like hobgoblins where you don't expect them, and vanish just as startlingly; or like repetition, which snatches us back to an earlier pattern or flings us forward as if on the crest of a wave. "Music is not just *in* time," Zuckerkandl writes. "It does something *with* time. . . . It is as if the even flow of time were cut up by the regularly recurrent sounds into short stretches of equal duration: the tones *mark time*." They stain time, then they reassemble it into small groups like so many lengths of cloth that have been dyed separately. At least our western music does; we're used to measured time in music. When polyphony came in, the only way it could make sense was if each of the voices kept the same time. But if we look back about 1,500 years or so, we find unmeasured time in music. A Gregorian chant, like poetry, simply improvised time. Even today, unless everyone used the same metronome it would be hard to agree on the right beat in measured time, so the beats agree with *one another*, not with an absolute. Ravel's mournful "Pavane for a Dead Princess" can sound lugubrious and heartrending when interpreted by one conductor, but almost sprightly by comparison when we hear a recording of it played by Ravel himself.

If you look at the interior of an early Romanesque church, say Saint-Étienne in Burgundy, which was built between 1083 and

---

*"Any space is as much a part of the instrument as the instrument itself."—Pauline Oliveros

1097, you find a massive architectural style with a high vaulted ceiling, parallel walls, and a long arcade—an ideal space for processions, but also for the reverberations of the Gregorian chant, which fills it like a dark wine poured into a heavy vessel. On the other hand, in a Gothic cathedral such as Notre Dame in Paris, with its nooks, corridors, statues, staircases, niches, and complex fugues in stone, a Gregorian chant would be broken up, fragmented. But at Saint-Étienne many voices can rise, mingle, and fill the elaborate space with glorious song.*

Western music has structures reminiscent of poetic verse forms. A sonata is as highly structured as the Malay verse form called a *pantoum*. The unstated warrant for the composer, as for the poet, is to stretch the limits of the form, to try to fly within the narrow corridors of a cage. That tension between the bright prison of a form and the freedom of imagination is what artistic genius is all about. Berlioz, for example, in his beautifully sensuous opera *Béatrice et Bénédict*, created music both grandiose and intimate. The duets shimmer with close, soulful harmony, the arias surge with an obsessive yearning that at some point breaks into melodic sobbing and sighing. It's an emotional ordeal that's personal and yet also larger than any one moment or heart. Zuckerkandl asks: "Who is man, that this almost-nothing, this 'nothing but tones' could become one of his most significant experiences?"

In the Argentinean film *Man Facing Southeast*, Rantes, an extraterrestrial playing an organ in the chapel at an insane asylum, says, "It's only a series of vibrations, but they have a good effect on the men. Where does the magic lie? In the instruments? In the one who wrote it? In me? In those that hear it? I cannot understand what they feel. Yes. I can understand. I just can't feel it." Later he explains that sensations upset the people of his planet, who can be destroyed by a catchy saxophone melody or a luscious perfume. He is not the only emissary from his planet sent to ours to investigate our one weapon against which they have no defense: human stupid-

---

*This very modern-sounding observation was also made by Abbé Suger, a counselor to Eleanor of Aquitaine, in the twelfth century.

ity. Sometimes the agents lose their way, become traitors, destroy themselves. A young, beautiful woman, Beatriz, who visits him in the asylum, we ultimately learn, is one of those lost agents who have become dangerously infatuated by the beauty of human sensory experience, unhinged by hearing a clarinet solo, "corrupted by sunsets, by certain fragrances . . ."

EARTH CALLING

We think of music as an invention, something that fulfills an inner longing, perhaps, to be an integral part of the sounds of nature. But not everyone perceives music in that way. About eighty miles north of Bangkok, in the foothills of Wat Tham Krabok, is a Buddhist temple where a group of concerned monks help drug addicts to recover. They use a combination of herbal therapy, counseling, and vocational training. One of the monks, Phra Charoen, a sixty-one-year-old naturalist by disposition, also busies himself in the music room, where, with electronic equipment, he records the electrical phenomena of the earth, which he then translates into musical notation. Charoen and his team of monks and nuns trace the fluctuating sound patterns onto transparent paper, then transfer the graphs to thin strips of cloth that can be catalogued and rolled up for storage. The graphs match up with the traditional eighteen-bar phrases of Thai music. These "pure melodies" are then played on a Thai instrument with an electronic organ as backup, and the result is recorded. Charoen's group are not musicians themselves, but they believe that music is not an imaginary thing, nor even something produced only by people; music falls out of the earth's rocks and roots, its trees and rain.* One western woman wrote that "under the temple trees, with birdsong filling the musical pauses, the visitor sits . . . and hears the earth of ancient Ayuthaya sing, or the stones of the Grand Palace, the sidewalks of Bangkok—or the cracks in the Hua Lampong Railway Station forecourt."

---

*In *The Heart of the Hunter,* Laurens van der Post reports that Bushmen speak of someone's death like this: "The sound which used to ring in the sky for him no longer rings."

This would no doubt strike a familiar chord with the American composer Charles Dodge, who, in June and September 1970, recorded "the sun playing on the magnetic field of the earth" by feeding magnetic data for 1961 into a specially programmed computer and synthesizer. The performance has a subtitle—"realizations in computed electronic sound"—and three "scientific associates" are prominently mentioned on the album's cover. The result is at times booming, at times squeaking, but consists mainly of shimmering, cascadingly melodic violin and woodwind sounds. Harmonious and breathy, they often create small flourishes and partial fanfares; they don't seem random at all, but rather energized by what, for lack of a better word, I'll call *entelechy*, that dynamic restlessness working purposefully toward a goal we associate with composed music. I also have a recording of Jupiter's magnetic field, a gift from the TRW corporation to visitors to the Jet Propulsion Laboratory during the encounters of Voyager I and II with Jupiter in 1980. An electric-field detector aboard the spacecraft recorded a stream of ions, the chirping of heated electrons, the vibrating of charged particles, lightning whistling across the planet's atmosphere, all accompanied by an aurora we hear as a hiss. Gas from a volcano on the moon Io adds a tinkling and a banshee-like scream of radio waves. Fascinating as this concert is, and useful to scientists, it doesn't sound like music, nor is it supposed to, but music could easily be woven from or around it. Artists have always looked to nature for their organic forms, and so it's not surprising to find a rather pop-sounding composition called "Pulsar." Over four hundred pulsars are known, at various distances from Earth. Using the recorded rhythmic pulses of once-massive stars about 15,000 light-years away, the composer offers Caribbean-like melodies, in which his "drummer from outer space," as he puts it, supplies percussion. The pulsars are identified on the record sleeve by number—083 − 45 on side one and 0329 + 54 on side two—as if they were indeed side men who sat in on the session. On another occasion, Susumu Ohno, a California geneticist, assigned a different note to each of the four chemical bases in DNA (*do* for cytosine, *re* and *mi* for adenine, *fa* and *sol* for guanine, and *la* and *ti* for thymine) and then played the somewhat limited-sounding result. Our cells vibrate; there is music in them, even if we don't hear it. Different animals hear some frequencies better than we do. Perhaps a mite, lost in the canyon of a crease of skin, hears our cells ringing like a mountain of wind chimes every time we move.

When the earth calls, it rumbles and thunders; it creaks. In towns like Moodus, Connecticut, swarms of small earthquakes rattle the residents for months on end. The seismic center of the quake storm is a very small area only a few hundred yards wide near the north end of town. I'm amazed there haven't been horror films about a devil's sinkhole, or some equal abomination. Ground grumblings of this sort are now called "Moodus noises," but long ago, when the Wangunk Indians chose the area for their powwows because it was there the earth spoke to them, they called the spot Machemoodus, which meant "place of noises," and their myths told how a god made the noises by blowing angrily into a cave. Cluster earthquakes can sound as light as corks popping or as relentless as cavalry charging. "Thunder underfoot" is how some have described it. "It's like you got hit on the bottom of your feet with a sledgehammer," one resident complains. The Moodus quakes are noisier than most because they're shallower (only about a mile deep; quakes along the San Andreas Fault are usually six to nine miles deep). Normal deep quakes lose much of their voice to the ground, which dampens and stills it. It may also be that the earth around Moodus simply conducts sound well. Since the town is located between two nuclear power plants, its residents grow anxious when the quakes rage for months, shifting and cracking the earth and sounding like a chronically rattling pantry.

At the Exploratorium in San Francisco, a pipe organ plays the sounds of San Francisco harbor as tide sloshes through its hollows, ringing with a thick brassy murmur. Now that the Russians and the Americans are planning a joint trip to Mars, I very much hope they'll take a set of panpipes along with them, so perfect for the windswept surface of Mars. Pipes would be an especially good choice because, although every culture on our planet makes music, each culture seems to invent drums and flutes before anything else. Something about the idea of breath or wind entering a piece of wood and filling it roundly with a vital cry—a sound—has captivated us for millennia. It's like the spirit of life playing through the whole length of a person's body. It's as if we could breathe into the trees and make them speak. We hold a branch in our hands, blow into it, and it groans, it sings.

Vision

*The greatest thing a human soul
ever does in this world is to see
something. . . . To see clearly
is poetry, prophecy, and
religion, all in one.*

John Ruskin,
Modern Painters

THE BEHOLDER'S EYE

Look in the mirror. The face that pins you with its double gaze reveals a chastening secret: You are looking into a predator's eyes. Most predators have eyes set right on the front of their heads, so they can use binocular vision to sight and track their prey. Our eyes have separate mechanisms that gather the light, pick out an important or novel image, focus it precisely, pinpoint it in space, and follow it; they work like top-flight stereoscopic binoculars. Prey, on the other hand, have eyes at the sides of their heads, because what they really need is peripheral vision, so they can tell when something is sneaking up behind them. Something like us. If it's "a jungle out there" in the wilds of the city, it may be partly because the streets are jammed with devout predators. Our instincts stay sharp, and, when necessary, we just decree one another prey and have done with it. Whole countries sometimes. Once we domesticated fire as if it were some beautiful temperamental animal; harnessing both its energy and its light, it became possible for us to cook food to make it easier to chew and digest, and, as we found out eventually, to kill germs. But we can eat cold food perfectly well, too, and did for thousands of years. What does it say about us that, even in refined dining rooms, our taste is for meat served at the temperature of a freshly killed antelope or warthog?

Though most of us don't hunt, our eyes are still the great monopolists of our senses. To taste or touch your enemy or your food, you have to be unnervingly close to it. To smell or hear it, you can risk being farther off. But vision can rush through the fields and up the mountains, travel across time, country, and parsecs of outer space, and collect bushel baskets of information as it goes. Animals that hear high frequencies better than we do—bats and dolphins, for instance—seem to see richly with their ears, hearing geographically, but for us the world becomes most densely informative, most luscious, when we take it in through our eyes. It may even be that abstract thinking evolved from our eyes' elaborate struggle to make sense of what they saw. Seventy percent of the body's sense receptors cluster in the eyes, and it is mainly through seeing the world that we appraise and understand it. Lovers close their eyes when they kiss because, if they didn't, there would be too many visual distractions to notice and analyze—the sudden close-up of the loved one's eyelashes and hair, the wallpaper, the clock face, the dust motes suspended in a shaft of sunlight. Lovers want to do serious touching, and not be disturbed. So they close their eyes as if asking two cherished relatives to leave the room.

Our language is steeped in visual imagery. In fact, whenever we compare one thing to another, as we constantly do (consider the country expression: "It was raining harder than a cow pissing sideways on a rock"), we are relying on our sense of vision to capture the action or the mood. Seeing is proof positive, we stubbornly insist ("I saw it with my own eyes . . ."). Of course, in these days of relativity, feats of magic, and tricks of perception, we know better than to trust everything we see (". . . a flying saucer landed on the freeway . . ."). See with our naked eyes, that is. As Dylan Thomas reminds us, there are many "fibs of vision."* If we extend our eyes by attaching artificial lenses and other accessories to our real ones

---

*Among the many fibs of vision are optical illusions. A puddle forms on the highway in front of you. But, unlike a real puddle, it keeps moving farther away as you approach it. Because it is a hot summer day, with a layer of hot air sitting below a layer of cold air, a reflection (of the sky) is cast onto the road. The word "mirage" slowly forms in your mind. Its etymology means "to wonder at." When we look at something red, the lens of our eye adjusts to the same shape it needs for seeing something green that is closer. When we look at something blue, the lens changes in the opposite direction. As a result, blue things appear to recede into the background, and red things seem to leap forward. Red things seem to be contracting, while blue ones seem to be spreading out. Blue things are thought to be "cold," while pink things are thought to be "warm." And because the eye is always trying to make sense of life, if it encounters a puzzling scene it corrects the picture to what it knows. If it finds a familiar pattern, it sticks to it, regardless of how inappropriate it might be in that landscape or against that background.

(glasses, telescopes, cameras, binoculars, scanning electron micro-scopes, CAT scans, X-rays, magnetic resonance imaging, ultrasound, radioisotope tracers, lasers, DNA sequencers, and so on), we trust the result a little more. But Missouri is still called the *Show Me!* state, which, as a kind of visual pun, I guess, it displays on its license plates for motorists to see. "The writing is on the wall," a politician says sagely, forgetting temporarily that it could be a forgery nonetheless. We quickly see through people whose characters are transparent. And, heaven knows, we yearn for enlightenment. "I see where you're coming from," one woman says to another in a café, "but you'd better watch out, he's bound to see what you're up to." *See for yourself!* the impatient exclaim to disbelievers. After the Bible's first imperative—"Let there be light"—God viewed each day's toil and "saw that it was good." Presumably, He, too, had to see it to believe it. Ideas dawn on us, if we're bright enough, not dim-witted, espe-cially if we're visionary. And, when we flirt, though the common phrase sounds quite ghoulish and extreme, we give someone the eye.

The process of seeing began very simply. In the ancient seas, life-forms developed faint patches of skin that were sensitive to light. They could then tell light from dark, and also the direction of the light source, but that was all. These skills turned out to be so useful that eyes evolved that could judge motion, then form, and finally a dazzling array of details and colors. One reminder of our oceanic origins is that our eyes must be constantly bathed in salt water. Some of the oldest eyes on record are those of the trilobite, one of the great success stories of the Cambrian age, which we now know only through its plentiful fossil remains. As I type this, I am wearing on a chain around my neck a small trilobite fossil, set in a silver bezel. Five hundred million years ago, it thrived in the swamps, with compound faceted eyes that could see mainly sideways but, unfortu-nately, not up. On the other hand, the newest eyes are those we have invented, such as the electric eye (based on what we learned about the motion-detecting design of the frog's eye), or the mirror tele-scope (based on the contrast-judging design of the horseshoe crab's eye), or synchronous lenses for use in microsurgery, optical scanning, and severe vision problems (based on the double lens of copilia, a myopic crustacean that lives deep in the Mediterranean). Although plants do not have eyes, Loren Eiseley argues eloquently for the eye of the fungus pilobolus, which has a light-sensitive area that controls the spore cannon it aims at the brightest spot it can find.

We think of our eyes as wise seers, but all the eye does is gather light. Let's consider the light-harvesting. As we know, the eye works a lot like a camera; or rather, we invented cameras that work like our eyes. To focus a camera, you move the lens closer to or farther away from an object. The eye's rubbery, bean-shaped crystalline lens achieves the same result by changing its shape—the lens thins to focus on a distant object, which looks small; thickens to focus on a near one, which looks large. A camera can control the amount of light it allows in. The iris of the eye, which is really a muscle, changes the size of a small hole, the pupil,* through which the light enters the eyeball. Because fish don't have this pupillary response, in which the iris protects against sudden surges of light, and most of them do not have eyelids (since their eyes are constantly bathed in water), they're much more susceptible to dazzlement than we are. In addition to its gate-keeping function, the iris, named after the Greek word for rainbow, is what gives our eyes their color. Caucasian eyes appear blue at birth, Negro eyes brown. After death, Caucasian eyes appear greenish-brown. Blue eyes are not inherently blue, not *stained* blue like fabric: They appear blue because they have less pigment than brown eyes. When light enters "blue" eyes, the very short blue light rays scatter as they jump off tiny, nonpigmented particles; what we see are the scattered rays, and the eyes appear to be blue. Dark eyes have densely packed pigment molecules and absorb the blue wavelengths, at the same time reflecting other colors whose rays are longer. They therefore appear to be brown or hazel. Though on casual inspection irises may look pretty much the same, the pattern of color, starbursts, spots, and other features is so highly

---

*From the Latin *pupilla,* "a little doll." When the Romans looked into one another's eyes, they saw a doll-like reflection of themselves. The old Hebrew expression for pupil is similar: *eshon ayin,* which means "little man of the eye."

individual that law-enforcement people have considered using iris patterns in addition to fingerprints.

At the back of a camera, film records the images. Lining the rear wall of the eyeball is a thin sheet, the retina, which includes two sorts of photosensitive cells, rods and cones. We need two because we live in the two worlds of darkness and light. A hundred and twenty-five million thin, straight rods construe the dimness, and report in black and white. Seven million plump cones examine the bright, color-packed day. There are three kinds of cones, specializing in blue, red, and green. Mixed together, the rods and cones allow the eye to respond quickly to a changing scene. One place on the retina, where the optic nerve enters the brain, has no rods or cones at all and, as a result, does not perceive light; we refer to it as our "blind spot." But right in the middle of the retina lies a small crater, the fovea, filled with highly concentrated cones, which we use for precision focusing when we want to examine an object in bright light, to drag it into sharp view and grip it with our eyes. Because the fovea is so small, it can perform its magic only on a small area (a four-inch-square snapshot at eight feet, for example). Almost every cone in a fovea has its own direct line to higher centers in the brain; elsewhere on the retina, rods and cones may serve many cells, and vision is vaguer. The eyeball moves subtly, continuously, to keep an object in front of the fovea. In dim light, the fovea's cones are almost useless; instead we must look just "off" of an object to see it clearly with the surrounding rods, not directly at it because the fovea would fail us and the object appear invisible. Because the rods see no color, we don't perceive color at night. When the retina observes something, neurons pass the word along to the brain through a series of electro-chemical handshakes. In about a tenth of a second, the message reaches the visual cortex, which begins to make sense of it.

However, seeing, as we think of it, doesn't happen in the eyes but in the brain. In one way, to see flamboyantly, in detail, we don't need the eyes at all. We often remember scenes from days or even years earlier, viewing them in our mind's eye, and can even picture completely imaginary events, if we wish. We see in surprising detail when we dream. Sometimes when I'm in a visually besotting land-scape, somewhere out in nature and experiencing intense rapture, I lie down at night and close my eyes, and see the landscape parading across the inside of my closed lids. The first time this happened—on a 200,000-acre working cattle ranch, surrounded by pastel mesas, in the New Mexico desert—I was a little spooked. Wrung out from the rigors of the branding corral, I needed sleep, but all the day's images, gestures, and motions still blazed in my visual memory. It was not like dreaming: it was like trying to sleep with your eyes wide open during a fiesta in full swing.

The same thing happened more recently, this time in Antarctica. One sunny day, we cruised through Gerlache Strait, which narrows to 1600 feet at its southern end; ice mountains towered on either side of the ship. Black jagged mountains, covered in cascading snow and ice, looked like penguins standing in familiar postures in a wash of brilliant light. While real penguins porpoised beside the boat, huge icebergs floated by, with bases of pale blue and sides of mint green. In the ship's glassed-in observation deck, people sat in arm-chairs at the window, some dozing. One man held out his pinky and first finger as if giving someone the evil eye, but he was measuring an iceberg. Deception Island, though distant, looked close and clear in the sterile air. A crib of ice holding a soft blue wash in its palms drifted close to the ship. Across the strait, ice calved off a glacier with a loud explosive crumble. Pastel icebergs roamed around us, some tens of thousands of years old. Great pressure can push the air bubbles out of the ice and compact it. Free of air bubbles, it reflects light differently, as blue. The waters shivered with the gooseflesh of small ice shards. Some icebergs glowed like dull peppermint in the sun—impurities trapped in the ice (phytoplankton and algae) tinted them green. Ethereal snow petrels flew around the peaks of the icebergs, while the sun shone through their translucent wings. White, silent, the birds seemed to be pieces of ice flying with purpose and grace. As they passed in front of an ice floe, they became invisible. Glare transformed the landscape with such force that it seemed like a pure color. When we went out in the inflatable motorized rafts called Zodiacs to tour the iceberg orchards, I grabbed a piece of glacial ice and held it to my ear, listening to the bubbles cracking and popping as the air trapped inside escaped. And that night, though exhausted from the day's spectacles and doings, I lay in my narrow bunk, awake with my eyes closed, while sunstruck icebergs drifted across the insides of my lids, and the Antarctic peninsula revealed itself slowly, mile by mile, in the small theater of my closed eyes.

Because the eye loves novelty and can get used to almost any scene, even one of horror, much of life can drift into the vague background of our attention. How easy it is to overlook the furry yellow comb inside the throat of an iris, or the tiny fangs of a staple, or the red forked tongue of a garter snake, or the way intense sorrow makes people bend their bodies as if they were blowing in a high wind. Both science and art have a habit of waking us up, turning on all the lights, grabbing us by the collar and saying *Would you please pay attention!* You wouldn't think something as complexly busy as life would be so easy to overlook. But, like supreme racehorses, full of vitality, determination, and heart, we tend to miss sights not directly in our path—the colorful crowds of people on either side, the shapes left in the thickly rutted track, and the permanent spectacle of the sky, that ever-present, ever-changing pageant overhead.

HOW TO WATCH THE SKY

I am sitting at the edge of the continent, at Point Reyes National Seashore, the peninsula north of San Francisco, where the land gives way to the thrall of the Pacific and the arching blue conundrum of the sky. When cricket-whine, loud as a buzz saw, abruptly quits, only bird calls map the quiet codes of daylight. A hawk leans into nothingness, peeling a layer of flight from thin air. At first it flaps hard to gain a little altitude, then finds a warm updraft and cups the air with its wings, spiraling up in tight circles as it eyes the ground below for rodents or rabbits. Banking a little wider, it turns slowly, a twirling parasol. The hawk knows instinctively that it will not fall. The sky is the one visual constant in all our lives, a complex backdrop to our every venture, thought, and emotion. Yet we tend to think of it as invisible—an absence, not a substance. Though we move through air's glassy fathoms, we rarely picture it as the thick heavy arena it is. We rarely wonder about the blue phantasm we call the sky. *"Skeu,"* I say out loud, the word that our ancient ancestors used; I try to utter it as they might have, with fear and wonder: *"Skeu."* Actually, it was their word for a covering of any sort. To them, the sky was a roof of changing colors. Small wonder they billeted their gods there, like so many quarrelsome neighbors who, in fits of temper, hurled lightning bolts instead of crockery.

Look at your feet. You are standing in the sky. When we think of the sky, we tend to look up, but the sky actually begins at the earth. We walk through it, yell into it, rake leaves, wash the dog, and drive cars in it. We breathe it deep within us. With every breath, we inhale millions of molecules of sky, heat them briefly, and then exhale them back into the world. At this moment, you are breathing some of the same molecules once breathed by Leonardo da Vinci, William Shakespeare, Anne Bradstreet, or Colette. Inhale deeply. Think of *The Tempest.* Air works the bellows of our lungs, and it powers our cells. We say "light as air," but there is nothing light-weight about our atmosphere, which weighs 5,000 trillion tons. Only a clench as stubborn as gravity's could hold it to the earth; otherwise it would simply float away and seep into the cornerless expanse of space.

Without thinking, we often speak of "an empty sky." But the sky is never empty. In a mere ounce of air, there are 1,000 billion trillion gyrating atoms made up of oxygen, nitrogen, and hydrogen, each a menagerie of electrons, quarks, and ghostly neutrinos. Sometimes we marvel at how "calm" the day is, or how "still" the night. Yet there is no stillness in the sky, or anywhere else where life and matter meet. The air is always vibrant and aglow, full of volatile gases, staggering spores, dust, viruses, fungi, and animals, all stirred by a skirling and relentless wind. There are active flyers like butterflies, birds, bats, and insects, who ply the air roads; and there are passive flyers like autumn leaves, pollen, or milkweed pods, which just float. Beginning at the earth and stretching up in all directions, the sky is the thick, twitching realm in which we live. When we say that our distant ancestors crawled out onto the land, we forget to add that they really moved from one ocean to another, from the upper fathoms of water to the deepest fathoms of air.

The prevailing winds here are from the west, as I can see from the weird and wonderful shapes of the vegetation along the beach. A light steady breeze blowing off the Pacific has swept back the wild grasses into a sort of pompadour. A little farther back, in a more protected glade, I find a small clump of them, around which a circle runs in the dirt. It looks as if someone pressed a cookie cutter down in the ground, but the wind alone has done it, blowing the grass around and turning it into a natural protractor. We think of the wind as a destructive force—a sudden funnel that pops a roof off a schoolhouse in Oklahoma—but the wind is also a gradual and powerful mason that carves cliffs, erodes hillsides, re-creates beaches, moves trees and rocks down mountains or across rivers. Wind creates waves, as in the sensuously rippling dunes of Death Valley or along the changing shorelines. The wind hauls away the topsoil as if it were nothing more than a dingy tablecloth on the checkerboard fields of the Midwest, creating a "dust bowl." It can power generators, gliders, windmills, kites, sailboats. It sows seeds and pollen. It sculpts the landscape. Along rugged coasts, one often sees trees dramatically carved by the relentless wind.

The north wind is shown on ancient maps as a plump-cheeked man with tousled hair and a strained expression, blowing as hard as he can. According to Homer, the god Aeolus lived in a palatial cave, where he kept the winds tied up in a leather bag. He gave the bag to Odysseus to power his ship, but when Odysseus's comrades opened the bag the winds raced free throughout the world, squabbling and whirling and generally wreaking havoc. "The children of morning," Hesiod called the Greek winds. To the ancient Chinese, *fung* meant both wind and breath, and there were many words for the wind's temperaments. *Tiu* meant "to move with the wind like a tree." *Yao* was the word for when something floated on the breeze like down. The names of winds are magical, and tell a lot about the many moods the sky can take. There's Portugal's hillside *vento coado;* Japan's demonic *tsumuji,* or soft pine-grove-loving *mat-*

sukaze; Australia's balmy *brickfielder* (which first described dust storms blowing off brickyards near Sydney); America's moist warm *chinook* drifting in from the sea, and named after the language of Indians who settled Oregon; or snow-clotted *blizzard,* or fierce *Santa Ana,* or Hawaii's humid *waimea;* North Africa's hot, sand-laden desert *simoom* (from the Aramaic word *samma,* "poison"); Argentina's baking, depleting *zonda,* which pours down from the Andes to sweep the pampas; the Nile's dark, gloomy *haboob;* Russia's gale-force *buran,* bringing a storm in the summer or a blizzard in the winter; Greece's refreshing summer *etesian;* Switzerland's warm, gusty *foehn* blowing off the leeward slopes of a mountain; France's dry cold *mistral* ("master wind") squalling through the Rhône Valley and down to the Mediterranean coast; India's notorious *monsoon,* whose very name means a whole season of monsoons; the Cape of Good Hope's *bull's-eye squall;* Alaska's petulent *williwaw;* Gibraltar's easterly-blowing *datoo;* Spain's mellifluous *solano;* the Caribbean's *hurricane* (derived from the Taino word *huracan,* which means "evil spirit"); Sweden's gale-level *frisk vind;* China's whispering *I tien tien fung,* or first autumn breeze, the *sz.*

Storms have been fretting the coast here for days, and now thick gray clouds stagger across this sky. I watch mashed-potato heaps of cumulus (a word that means "pile") and broad bands of stratus (which means "stretched out"). As author James Trefil once observed, a cloud is a sort of floating lake. When rising warm air collides with descending cold air, the water falls, as it does now. I take shelter on a porch, while a real toad-strangler starts, a full-blooded, hell-for-leather thunderstorm, during which the sky crackles and throbs. Lightning appears to plunge out of it, a pitchfork stabbing into the ground. In fact, it sends down a short electrical scout first, and the earth replies by arcing a long bolt up toward the sky, heating the air so fast that it explodes into a shock wave, or *thunder,* as we call it. Counting the seconds between a lightning flash and the thunder, I then divide by five, and get a rough idea of how far away it is—seven miles. In one second, sound travels 1,100 feet. If the lightning flash and the thunder arrive at the same time, one doesn't have much of a chance to count. In a little while the storm quiets, as the thunder bumpers roll farther up the coast. But some clouds still stalk the sky. A cloud rhinoceros metamorphoses into a profile of Eleanor Roosevelt; then a bowl of pumpkins; then a tongue-wagging dragon. Parading hugely across the sky, clouds like these have squatted above people of all times and countries. How many vacant afternoons people have passed watching the clouds drift by. The ancient Chinese amused themselves by finding shapes in the clouds just as Inuits, Bantus, and Pittsburghers do now. Sailors, generals, farmers, ranchers, and others have always consulted the crystal ball of the sky to foretell the weather (lens-shaped clouds—severe winds aloft; dappled or "mackerel" sky—rain is near; low, thick, dark, blanketlike clouds—a stormy cold front may be coming), devising jingles, maxims, and elaborate cloud charts and atlases, graphics as beautiful as they are useful. On a train through Siberia, Laurens van der Post looked out the window at the huge expanse of flat country and endless sky. "I thought I had never been to any place with so much sky and space around it," he writes in *Journey into Russia*, and was especially startled by "the immense thunder clouds moving out of the dark towards the sleeping city resembling, in the spasmodic lightning, fabulous swans beating towards us on hissing wings of fire." As van der Post watched the lightning from the train, the Russian friend accompanying him explained that they had a special word in his language for just that scene: *Zarnitsa*.

Throughout time and place, people have been obsessed with the many moods of the sky. Not just because their crops and journeys depended on the weather, but because the sky is such a powerful symbol. The sky that gods inhabit, the sky whose permanence we depend on and take for granted, as if it really were a solid, vaulted ceiling on which stars were painted, as our ancestors thought. The sky that can fall in nursery rhymes. In the nuclear disarmament marches of the sixties, some people wore signs that read: CHICKEN LITTLE WAS RIGHT. We picture the sky as the final resting place of those we love, as if their souls were perfumed aerosol. We bury them among pine needles and worms, but in our imaginations we give them a lighter-than-air journey into some recess of the sky from which they will watch over us. "High" is where lofty sentiments dwell, where the "high and mighty" live, where choirs of angels sing. I don't know why the sky symbolizes our finest ideals and motives, unless, lacking in self-confidence, we think our acts of mercy, generosity, and heroism are not intrinsic qualities, not characteristics human beings alone can muster, but temporary gifts from some otherworldly power situated in the sky. Stymied by events, or appalled by human nature, we sometimes roll our eyes upward, to where we believe our fate is dished out in the mansions of the stars.

Driving four hours south, along spectacular cliffs and a wild and dramatic ocean where sea otters bob in the kelp beds, sea lions bark, harbor seals clump together like small mountain ranges, and pelagic cormorants, sanderlings, murres, and other seabirds busily nest, I pause on a wind-ripped slope of Big Sur. A Monterey pine leans out over the Pacific, making a ledge for the sunset. The pummeling gales have strangled its twigs and branches on the upwind side, and it looks like a shaggy black finger pointing out to sea. People pull up in cars, get out, stand and stare. Nothing need be said. We all understand the visual nourishment we share. We nod to one another. The cottony blue sky and dark-blue sea meet at a line sharp as a razor's edge. Why is it so thrilling to see a tree hold pieces of sky in its branches, and hear waves crash against a rocky shore, blowing spray high into the air, as the seagulls creak? Of the many ways to watch the sky, one of the most familiar is through the filigree limbs of a tree, or around and above trees; this has much to do with how we actually see and observe the sky. Trees conduct the eye from the ground up to the heavens, link the detailed temporariness of life with the bulging blue abstraction overhead. In Norse legend, the huge ash tree Yggdrasil, with its great arching limbs and three swarming roots, stretched high into the sky, holding the universe together, connecting earth to both heaven and hell. Mythical animals and demons dwelt in the tree; at one of its roots lay the well of Mimir, the source of all wisdom, from which the god Odin drank in order to become wise, even though it cost him the loss of an eye. We find trees offering us knowledge in many of the ancient stories and legends, perhaps because they alone seem to unite the earth and the sky—the known, invadable world with everything that is beyond our grasp and our power.

Today the ocean pours darkly, with a white surf pounding over and over. Close to the shore, the thick white wave-spume looks applied by a palette knife. The damp, salty wind rustles like taffeta petticoats. One gull finds a shellfish and begins picking it apart, while the others fly after it and try to snatch the food away, all of them squeaking like badly oiled machinery.

When I was in Istanbul many years ago, I marveled at the way the onion-shaped mosques carved the sky between them. Instead of seeing a skyline, as one would in New York or San Francisco, one saw only the negative space between the swirling, swooping, spiraling minarets and bulbous domes. But here one sees the silhouette of distinctive trees against the sky: Scotch pine, which has a long stem with a roundish top resembling a child's rattle; tall, even, rice-grain-shaped cypress and spruce. Farther north stand the sequoias, the heaviest living things to inhabit the planet. The talcy-leaved eucalyptus, nonnative trees that are so hardy and fast-growing they've taken over whole forests in California, look like bedraggled heads of freshly shampooed hair. In the fall and winter, one can find among their branches long garlands of monarch butterflies, hanging on by their feet, which have prongs like grappling hooks. Each year, a hundred million migrate as much as four thousand miles from the northern United States and Canada to overwinter on the California coast. They cluster to keep warm. Butterflies seem to prefer the oily mentholated groves, the fumes of which keep away most insects and birds. Blue jays occasionally attack the monarchs when they leave their garland to sip nectar or sit out in the open and spread their wings wide as solar collectors. Monarch larvae eat the leaves of milkweed, a poisonous, digitalislike plant, to which they are immune, but which makes *them* poisonous; and birds quickly learn that eating monarchs will make them sick. If you see a monarch flying around with a wedge-shaped piece of wing missing, you are most likely looking at a veteran of an uninformed bird's attack. When I was helping to tag monarchs, I saw just such a female trembling on the porch floor outside my motel-room window. A huge blue jay in a nasty temper perched on the porch rail, screeching and flapping, and getting ready to dive at the monarch again. Though I usually know better than to intrude in nature's doings, my instincts took over and I rushed outside, lunged at the blue jay to punch it in the chest, just as it leapt up with a great squawk and flap, truly terrified by my sudden attack. The butterfly stood her ground and shook, and I picked her up carefully, checked to see if she were pregnant by pressing her abdomen gently between my thumb and forefinger, feeling for a hard pellet. She wasn't, and the missing wedge of wing didn't look too bad, so I carried her to the base of a tree, at the top of which swayed a long orange string of monarchs. Then I held her above my open mouth and breathed warm air over her body, to help heat her flying muscles since it was a chilly morning, and tossed her into the air. She fluttered right up to her cluster, and, as I walked back to my room, I saluted her. The blue jay was still shrieking bloody murder, and then I saw it fly out of the yard with strong, confident beats.

At Big Sur, the hawks are working the thermals like barnstormers, swooping and banking as they ride invisible towers of warm, rising air above the sun-heated ground. Birds are so nimble and adroit. Each species has its own architecture, flight habits, and talents to make the most of the sky, which they sometimes reveal in their silhouettes. On some owls, for instance, the leading edge of the primary feathers is softly fringed to muffle the sound of their approach. Finches flap hard a few beats, then close their wings and rest a little. Turtledoves flap continuously when they're flying. Peregrine falcons fold in their wings when they dive. Swifts, which average about twenty-five mph, have very pointy wings that make them sleeker by cutting down on drag as they dart and glide. At the Grand Canyon, you can see them working the canyon walls like small aerobats.

Our sky is also filled with "passive flyers." Female ash trees loose their winged "keys," and aspens and others produce long catkins that drop and blizzard across the ground. Maples launch tadpole-shaped seeds that fall whirlygig down, all blade, all propeller, like small autogyros. Thanks to the wind, the sex lives of many plants have changed. Dandelions, milkweed, thistles, cottonwoods, and others have evolved wind-riders in the shape of parachutes or sails. Pine, spruce, hemlock, maple, oak, and ragweed don't have flamboyant flowers, but they don't need them to divert a bird or bee. The wind is go-between enough. Plants can't court, or run away from a threat, so they've devised ingenious ways to exploit their environment and animals. Pollen grains may be as small as one ten-thousandth of an inch in diameter, yet they must travel uncertain winds and strike home. Using a wind tunnel, Karl Niklas, a Cornell scientist, recently discovered that plants aren't just hobos, hoping their pollen will catch a passing breeze and get off at the right stop. Niklas found that the pine cone has evolved an architecture perfect for capturing wind from any direction: a turbine shape, with petal-blades that spin the air all around it. Like a planet, the pine cone wraps itself in an atmosphere of rapidly moving air, with, just below the upper, swirling layer, a still and vacant layer. When pollen falls from the rapid layer to the still layer, it cascades right down into the cone. Niklas also tested the air-flow dynamics of the jojoba plant, which uses two rabbit-ear-shaped leaves to direct air, with results that show similar finesse.

In allergy season, pollen makes me (and millions of others) sneeze a little, and my eyes sometimes itch so that I can't wear my contact lenses. But I like knowing that all this mischief happens just because of shape. Tiny Sputniks traveling through the lower sky, some pollen looks like balls covered with spikes. Others are as football-shaped as the pupils of alligators. Pine pollen is round, with what looks like a pair of ears attached to each side. Their shapes make them move or fly at different speeds and in different patterns, and there's little danger of the wrong pollen swamping the wrong plant. It's odd to think of the sky having niches, but it does; even the wind has niches.

As night falls on Big Sur, all the soot of the world seems to pour down into the sunset. A swollen yellow doubloon drops slowly into the ocean, shimmer by shimmer, as if swallowed whole. Then, at the horizon, a tiny green ingot hovers for a second, and vanishes. The "green flash" people call it, with mystical solemnity. But it is the briefest flash of green, and this is the first time in all my sunset-watching that I've seen it. Green, azure, purple, red: How lucky we are to live on a planet with colored skies. Why is the sky blue? The sun's white light is really a bouquet of colored rays, which we classify into a spectrum of six colors. When white light collides with atoms of gases that make up the atmosphere—primarily oxygen and nitrogen—as well as with dust particles and moisture in the air, blue light, the most energetic light of the visible spectrum, is scattered. The sky seems to be full of blue. This is particularly true when the sun is overhead, because the light rays have a shorter distance to travel. The red rays are longer, and penetrate the atmosphere better. By the time the sun sets, one side of the Earth is turning away from the sun; the light has to travel farther, at an angle, through even more dust, water vapor, and air molecules; the blue rays scatter even more and the red rays remain, still traveling. The sun may appear magnified into a swollen ghost, or slightly elliptical, or even above the horizon when it's really below it, thanks to refraction, the bending of light waves. What we see is a glorious red sunset, especially if prowling clouds reflect the changing colors. The last color that plows through the atmosphere without being scattered is green, so sometimes we see a green flash right after the sun disappears. In space, the air appears to be black because there is no dust to scatter the blue light.

At Big Sur lighthouse, perched on a distant promontory, a beacon flashes to warn ships away from the coast and sandbanks, its light zooming out to them at 186,000 miles per second. The searchlight of the sun takes about eight minutes to reach Earth. And the light we see from the North Star set sail in the days of Shakespeare. Just think how straight the path of light is. Pass sunlight through a prism, though, and the light bends. Because each ray bends a different amount, the colors separate into a band. Many things catch the light prismatically—fish scales, the mother-of-pearl inside a limpet shell, oil on a slippery road, a dragonfly's wings, opals, soap bubbles, peacock feathers, the grooves in gramophone records, metal that's lightly tarnished, the neck of a hummingbird, the wing cases of beetles, spiders' webs smeared with dew—but perhaps the best known is water vapor. When it's raining but the sun is shining, or at a misty waterfall, sunlight hits the prismlike drops of water and is split into what we call a "rainbow." On such a day, rainbows are always about, hidden somewhere behind the skirts of the rain; but to see one best, you have to be positioned just right, with the sun behind you and low in the sky.

It is nighttime on the planet Earth. But that is only a whim of nature, a result of our planet rolling in space at 1,000 miles per minute. What we call "night" is the time we spend facing the secret reaches of space, where other solar systems and, perhaps, other planetarians dwell. Don't think of night as the absence of day; think of it as a kind of freedom. Turned away from our sun, we see the dawning of far-flung galaxies. We are no longer sun-blind to the star-coated universe we inhabit. The endless black, which seems to stretch forever between the stars and even backwards in time to the Big Bang, we call "infinity," from the French *in-fini*, meaning unfinished or incomplete. Night is a shadow world. The only shadows we see at night are cast by the moonlight, or by artificial light, but night itself is a shadow.

In the country, you can see more stars, and the night looks like an upside-down well that deepens forever. If you're patient and wait until your eyes adjust to the darkness, you can see the Milky Way as a creamy smudge across the sky. Just as different cultures have connected the stars into different constellations, they've seen their own private dramas in the Milky Way. The "backbone of night" the Bushmen of the Kalahari call it. To the Swedes, it is the "winter street" leading to heaven. To the Hebridean islanders, the "pathway of the secret people." To the Norse, the "path of ghosts." To the Patagonians, obsessed with their flightless birds, "the White pampas where ghosts hunt rheas." But in the city you can see the major constellations more easily because there are fewer stars visible to distract you.

Wherever you are, the best way to watch stars is lying on your back. Tonight the half-moon has a Mayan profile. It looks luminous and shimmery, a true beacon in the night, and yet I know its brilliance is all borrowed light. By day, if I held a mirror and bounced a spot of sunlight around the trees, I would be mimicking how the moon reflects light, having none of its own to give. Above me, between Sagittarius and Aquarius, the constellation Capricorn ambles across the sky. The Aztecs pictured it as a whale *(cipactli)*, the East Indians saw an antelope *(makaram)*, the Greeks labeled it "the gate of the gods," and to the Assyrians it was a goat-fish *(munaxa)*. Perhaps the best-known star in the world is the North Star, or Polaris, though of course it has many other names; to the Navaho, it is "The Star That Does Not Move," to the Chinese, the "Great Imperial Ruler of Heaven."

Throughout time, people have looked up at the sky to figure out where they were. When I was a girl, I used to take an empty can, stretch a piece of tinfoil over one end and pierce pinholes in it in the outline of a constellation; then I'd shine a flashlight in the other end, and have my own private planetarium. How many wanderers, lost on land or sea, have waited till night to try and chart their way home with help from the North Star. Locating it as they did connects us across time to those early nomads. First you find the Big Dipper and extend a line through the outer two stars of its ladle. Then you'll see that the North Star looks like a dollop of cream fallen from the upside-down Dipper. If the Big Dipper isn't visible, you can find the North Star by looking for Cassiopeia, a constellation just below Polaris that's shaped like a W or an M, depending on the time you see it. To me, it usually looks like a butterfly. Because the Earth revolves, the stars seem to drift from east to west across the sky, so another way to tell direction is to keep your eye on one bright star in particular; if it appears to rise, then you're facing east. If it seems to be falling, you're facing west. When I was a Girl Scout, we found our direction during the day by putting a straight stick in the ground. Then we'd go about our business for a few hours and return when the stick cast a shadow about six inches long. The sun would have moved west, and the shadow would be pointing east. Sometimes we used a wristwatch as a compass: Place the watch face up, with the hour hand pointing toward the sun. Pick up a pine needle or twig and hold it upright at the edge of the dial so that it casts a shadow along the hour hand. South will be halfway between the hour hand and twelve o'clock. There are many other ways to tell direction, of course, since roaming is one of the things human beings love to do best—but only if they can count on getting home safely. If you see a tree standing out in the open, with heavy moss on one side, that side is probably north, since moss grows heaviest on the shadiest side of a tree. If you see a tree stump, its rings will probably be thicker on the sunny side, or south. You can also look up at the tops of pine trees, which mainly point east. Or, if you happen to know where the prevailing wind is coming from, you can read direction from the wind-bent grasses.

It's November. The Leonids are due in Leo. Pieces of comet that fall mainly after sunset or before sunrise, they appear in the same constellations each year at the same time. In Antarctica, I had hoped to see auroras, veils of light caused by the solar wind bumping into the earth's magnetic field and leaving a gorgeous shimmer behind. But our days were mainly sun-perfect, and our nights a grisly gray twilight. In the evening, the sea looked like pounded gunmetal, but there were no auroras to make glitter paths overhead. Here is how Captain Robert Scott described one display in June 1911:

> The eastern sky was massed with swaying auroral light . . . fold on fold the arches and curtains of vibrating luminosity rose and spread across the sky, to slowly fade and yet again spring to glowing life.
> The brighter light seemed to flow, now to mass itself in wreathing folds in one quarter, from which lustrous streamers shot upward, and anon to run in waves through the system of some dimmer figure. . . .
> It is impossible to witness such a beautiful phenomenon without a sense of awe, and yet this sentiment is not inspired by its brilliancy but rather by its delicacy in light and colour, its transparency, and above all by its tremulous evanescence of form.

Tonight Mars glows like a steady red ember. Though only a dot of light in the sky, it is in my mind a place of blustery plains, volcanoes, rift valleys, sand dunes, wind-carved arches, dry river beds, and brilliant white polar caps that wax and wane with the seasons. There may even have been a climate there once, and running water. Soon Venus will appear as a bright silvery light, as it usually does about three hours after sunset or before sunrise. With its gauzy white face, it looks mummified in photos, but I know that impression is given by cloud banks full of acids floating above a surface where tricks of light abound and the temperatures are hot enough to melt lead. There are many kinds of vision—literal, imaginative, hallucinatory; visions of greatness or of great possibilities. Although I can't see the steady light of other planets just yet, I know they are there all the same, along with the asteroids, comets, distant galaxies, neutron stars, black holes, and other phantoms of deep space. And I picture them with a surety Walt Whitman understood when he proclaimed: "The bright suns I see and the dark suns I cannot see are in their place."

Sunrise. Darkness begins to wash out of the sky. A thick lager of fog sits in the valley like the chrysalis of a moth. Venus, Mercury, and Saturn burn bright silver holes in the slowly bluing sky. The stars have vanished, because by the time starlight gets to Earth it's too dim to be seen during the daylight. Two black shapes in the fog reel into focus as cows. A calf reveals itself. Learning about the world is like this—watching and waiting for shapes to reveal themselves in the fog of our experience. A wan sky curdles with gauzy streaks of cloud. The land is veiled in mist. The highest hill looks like a train's smokestack: Clouds trail behind it. Now the cloud world that was horizontal becomes vertical as cumulus begin to rise over the mountain. Venus throbs, a broken lighthouse in the western sky. A nation of cloud tepees rises along the top of the ridge. The first hawk of the day glides on cool air, wings arched. The dew sits in round, bluish drops on the clover-rich grass. A squadron of eighteen pelicans flies in a long check mark overhead, turns on edge and vanishes, turns again and tilts back into sight. A huge pillow of fog rolls through the valley. The cows disappear, but the sky grows bluer; Venus fades, white clouds begin to form, the fog lifts like a fever, a house and more cows appear. A lone, lightning-struck tree stands like a totem pole on a hillside, the light quickens, and birds begin their earnest songs, as the first yellow floats up like egg yolk over the ledge of the world, and then the sun is a canary singing light.

LIGHT

Without light, could you or I see? Without light and water, could life exist at all? It's hard to imagine living without light. The most frightening dark I remember was when scuba-diving in an underwater cave in the Bahamas. We carried flashlights, but at one point I turned mine off and just sat in the darkness. Later, when I climbed up out of the cave and stepped into the blinding light of a hot Bahamian day, the sun was burning from ninety-three million miles away, yet felt like fresh sandpaper on my arms and legs. At exactly 4:00 P.M. it rained briefly, as it did each day at that time. The wet roads looked shiny. Not so the stone walls. Light waves hitting a smooth, flat surface bounce back evenly, making the surface shine. If the surface is rough, the light waves scatter in different directions, not as many will return to our eyes, and the surface doesn't look shiny. It takes only a little light to stimulate the eye—a candle burning ten miles away will do—and a moonlit night, especially after a snowfall, will flood the eye with reflections, shapes, and motion. Astronauts in orbit around Earth can see beneath them the wakes ships leave in the oceans. But when we're in the forest under a low cloud cover, and night falls like a black sledgehammer, there are no light rays to bounce back at the eyes, and we don't see. As Sir Francis Bacon noted slyly in his essay on religion, "All colours will agree in the dark."

Even people who have been blind since birth are greatly affected by light, because, although we need light to see, light also influences us in other subtle ways. It affects our moods, it rallies our hormones, it triggers our circadian rhythms. During the season of darkness in northern latitudes, the suicide rate soars, insanity looms in many households, and alcoholism becomes rampant. Some diseases, including rickets, result in part from children receiving too little sunlight; children are active creatures, and need the vitamin D that light produces to keep them healthy. Other malaises, like Seasonal Affective Disorder (SAD), which leaves many people feeling depleted and depressed in the winter months, can be corrected by daily doses of very bright light (twenty times brighter than average indoor lighting) for about half an hour each morning. Some lingering low-level depression can be cured by changing a patient's sleep schedule so that it parallels the season's periods of light and dark more closely. Most years, Ithaca, New York, has only two seasons, both of which are wet—hot wet and cold wet—so it tends to be overcast much of the time. Bright light doesn't stream in through the picture windows at sunrise. Anyway, my bedroom windows are thickly curtained, and I sleep in a room dark enough to please a star-nosed mole. Although I go speed walking for fifty minutes every day, regardless of season or weather, I find that I feel much more energetic, and generally happier, if I do my winter walkabouts in early or midmorning, and do them every single day without fail; in summer, it doesn't seem to matter when I work out, or even if I occasionally miss a day.

Light therapy is being used to help people with psoriasis, schizophrenia, and even some forms of cancer. The pineal gland, or "third eye," as it's been mystically labeled, seems to be intimately involved with our sense of season, of well-being, the onset of puberty, the amount of testosterone or estrogen we produce, and certain of our more subtle seasonal behaviors. Testosterone is at its highest in men during early afternoons (around 2:00 P.M.) in October, I suppose because a child conceived then would be born during the summer and have a greater chance of survival. Of course, men don't all wait for that one climactic autumn month to make love, rising through a crescendo of libido in September and an only slightly dwindling mania as they near Christmas.

One of the hallmarks of our species is our ability not only to adapt to our environment, but also to change the environment to better suit us. We withstand the cold reasonably well, but we don't let its extremes bully us into migrating; we just build shelters and wear clothes. We respond to sunlight, and we create light for times when there is little or no sun. We use the energy of fire, and we create energy. Most of this we like to do outside our bodies, unlike other creatures. When we want to light up the world around us, we build lamps. Many insects, fish, crustaceans, squids, fungi, bacteria, and protozoa bioluminesce: They throb with light. The angler fish even hangs a glowing lure from its mouth, which attracts prey. A male firefly flashes its cool, yellow-green semaphores of desire, and if the female, too, is randy, she flashes back her consent. They look hot and bothered, twinkling through a summer's night like lovers drifting from one streetlamp to the next. Their light comes from the blending of two chemicals, luciferin and luciferase (lucifer means "shining"). If you row through Phosphorescent Bay off the southwestern coast of Puerto Rico, at night, you'll leave a trail of glowing auroras in the water and see cool fire dripping from your oars; it comes from microscopic invertebrates, which live in the water and secrete a luminous fluid whenever jostled. James Morin, a marine biologist at UCLA, has been studying rice-grain-sized crustaceans of the genus *Vargula*, which he's nicknamed "firefleas." There are thirty-nine known species, and they use light not only for courtship, but also to alarm their enemies. When they light up, they become more visible, but so does the predator, which in turn becomes easier to spot by an even larger predator. During courtship, each species flashes its own dialect of light. Far brighter than fireflies, *Vargulae* glow with an intense brilliance. "If I put a single fireflea on my fingertip and squashed it, I could read a newspaper from the light for about ten minutes," Morin explains. Sailors tell about ships trailing fire from their sterns. They don't mean St. Elmo's fire (an atmospheric phenomenon that can strike a mast and ignite it with a cool, crackling, eerie green glow), but a moon-bright glitter swirled up on the water as the ship passes through tiny luminous lives.

Around Halloween, stores begin to sell necklaces, wands and other plastic items that glow coolly in the dark. Based on bioluminescence, they contain luciferins, and work the same way as a firefly's glow. But, for extra sparkle, a trick or treater might also chew wintergreen Lifesavers. If you stand in the dark and crush one between your teeth, it will spill blue-green flashes of light. Certain substances (some quartzes and mica, even adhesive tape, when it is yanked off specific surfaces) are *triboluminescent;* they give off light if you rub, crush, or break them. Broken wintergreen fluoresces and broken sugar gives off ultraviolet light; the combination—in candies that contain both sugar and oil of wintergreen—produces tiny bolts of blue-green lightning. Try this parlor game: Step into a closet with a mouthful of wintergreen Lifesavers and a friend and wait for sparks to fly.

COLOR

At twilight, pink wings tremble along the hilltops, and purple does a shadow dance over the lake. When light hits a red car on the streetcorner, only the red rays are reflected into our eyes, and we say "red." The other rays are absorbed by the car's paint job. When light hits a blue mailbox, the blue is reflected, and we say "blue." The color we see is always the one being reflected, the one that doesn't stay put and get absorbed. We see the rejected color, and say "an apple is red." But in truth an apple is everything *but* red.

Even though it's sunset and the quantity, quality, and brightness of light have all diminished, we still perceive the blue mailbox as blue, the red car as red. We are not really cameras. Our eyes do not just measure wavelengths of light. As Edwin Land, inventor of the Polaroid Land Camera and instant photography, deduced, we judge colors by the company they keep. We compare them to one another, and revise according to the time of day, light source, memory.* Otherwise, our ancestors wouldn't have been able to find food at sunset or on overcast days. The eye works with ratios of color, not with absolutes. Land was not a biologist, but a keen observer of how we observe, and his theory of color constancy, proposed in 1963, continues to make sense. Every college student at one time or another has asked what it means to *know* something, and whether there are simple perceptual truths that people share. We watch *color* television because our ancestors had eyes cued to the ripening of fruit; and they also had to be wary of poisonous plants and animals (which tend to be brightly colored). Most people can identify between 150 and 200 colors. But we do not all see exactly the same

---

*Because albinos lack a dark layer of cells behind the retina, more light travels around inside their eyes and colors often seem to them quieter and more diluted.

colors, especially if we're partly or completely color-blind,* as many people are—men in particular. A blue ship may not look the same when viewed from opposite sides of a river, depending on the landscape, clouds, and other phenemona. The emotions and memories we associate with certain colors also stain the world we see. And yet, how astonishing it is that we do tend to agree on what we call red or teal or cream.

Not all languages name all colors. Japanese only recently included a word for "blue." In past ages, *aoi* was an umbrella word that stood for the range of colors from green and blue to violet. Primitive languages first develop words for black and white, then add red, then yellow and green; many lump blue and green together, and some don't bother distinguishing between other colors of the spectrum. Because ancient Greek had very few color words, a lot of brisk scholarly debate has centered around what Homer meant by such metaphors as the "wine-dark sea." Welsh uses the word *glas* to describe the color of a mountain lake, which might in fact be blue, gray, or green. In Swahili, *nyakundu* could mean brown, yellow, or red. The Jalé tribespeople of New Guinea, having no word for green, are content to refer to a leaf as dark or light. Though English sports a fair range of words to distinguish blue from green (including azure, aqua, teal, navy, emerald, indigo, olive), we frequently argue about whether a color really should be *considered* blue or green, and mainly resort to similes such as grass green, or pea green. The color language of English truly stumbles when it comes to life's processes. We need to follow the example of the Maori of New Zealand, who have many words for red—all the reds that surge and pale as fruits and flowers develop, as blood flows and dries. We need to boost our range of greens to describe the almost squash-yellow green of late winter grass, the achingly fluorescent green of the leaves of high summer, and all the whims of chlorophyll in between. We need words for the many colors of clouds, surging from pearly pink during a calm sunset

---

*Oliver Sachs tells of a sixty-five-year-old artist who survived a car accident only to discover that his color vision had entirely vanished because of a brain injury. Human flesh appeared "rat-colored" to him, and he found food ghastly and inedible without color.

over the ocean to the electric gray-green of tornadoes. We need to rejuvenate our brown words for all the complexions of bark. And we need cooperative words to help refine colors, which change when they're hit by glare, rinsed with artificial light, saturated with pure pigment, or gently bathed in moonlight. An apple remains red in our minds, wherever we see it, but think how different its red looks under fluorescent light, on the shady branch of a tree, on a patio at night, or in a knapsack.

Color doesn't occur in the world, but in the mind. Remember the old paradoxical question: If a tree falls in the forest, and no one is around to hear it, does it make a sound? A parallel question in vision: If no human eye is around to view it, is an apple really red? The answer is no, not red in the way we mean red. Other animals perceive colors differently than we do, depending on their chemistry. Many see in black and white. Some respond to colors invisible to us. But the many ways in which we enjoy color, identify it, and use it to make life more meaningful are unique to humans.

In the Hall of Gems at the Museum of Natural History in New York, I once stood in front of a huge piece of sulfur so yellow I began to cry. I wasn't in the least bit unhappy. Quite the opposite; I felt a rush of pleasure and excitement. The intensity of the color affected my nervous system. At the time, I called the emotion wonder, and thought: Isn't it extraordinary to be alive on a planet where there are yellows such as this? One of today's "color consultants" might tell me instead which chakra, or energy center, the yellow was stimulating. The therapeutic use of color has become faddish of late, and, for a price, all sorts of people will help you "learn what colors your body needs," as one guru puts it. Recent books decree the only and perfect colors to make you look beautiful or cure your flagging spirits. But scientists have known for years that certain colors trigger an emotional response in people. Children will use dark colors to express their sadness when they're painting, bright colors to express happiness. A room painted bubble-gum pink (known in hospitals, schools and other institutions as "passive pink") will quiet them if they've gotten obstreperous. In a study done at the University of Texas, subjects watched colored lights as their hand-grip strength was measured. When they looked at red light, which excites the brain, their grip became 13.5 percent stronger. In another study, when hospital patients with tremors watched blue light, which calms the brain, their tremors lessened. Ancient cultures (Greek, Egyptian, Chinese, Indian, and others) used color therapies of many sorts, prescribing colors for various distresses of the body and soul. Colors can alarm, excite, calm, uplift. Waiting rooms in television studios and theaters have come to be called greenrooms, and are painted green because the color has a restful effect. Dressing baby boys in blue and girls in pink has a long history. To the ancients, a baby boy was cause for celebration, since it meant another strong worker and the carrying on of the family name. Blue, the color of the sky where the gods and fates lived, held special powers to energize and ward off evil, so baby boys were dressed in blue to protect them. Later, a European legend claimed that baby girls were born inside delicate pink roses, and pink became their color.

Some years ago, when I had taken a job directing a writing program in St. Louis, Missouri, I often used color as a tonic. Regardless of the oasis-eyed student in my office, or the last itchlike whim of the secretary, or the fumings of the hysterically anxious chairman, I tried to arrive home at around the same time every evening, to watch the sunset from the large picture window in my living room, which overlooked Forest Park. Each night the sunset surged with purple pampas-grass plumes, and shot fuchsia rockets into the pink sky, then deepened through folded layers of peacock green to all the blues of India and a black across which clouds sometimes churned like alabaster dolls. The visual opium of the sunset was what I craved. Once, while eating a shrimp-and-avocado salad at the self-consciously stately faculty club, while I gossiped with an anorexic and hopped-up young colleague, I found myself restless for the day to be over and all such tomblike encounters to pale, so I could drag my dinette-set chair up to the window and purge my senses with the pure color and visual tumult of the sunset. This happened again the next day in the coffee room, where I stood chatting with one of the literary historians, who always wore the drabbest camouflage colors and continued talking long after a point had been made. I set my facial muscles at "listening raptly," as she chuntered on about her specialty, the Caroline poets, but in my mind the sun was just beginning to set, a green glow was giving way to streaks of sulfur yellow, and a purple cloud train had begun staggering across the horizon. I was paying too much rent for my apartment, she explained. True, the apartment overlooked the park's changing seasons, had a picture window that captured the sunset every night, and was only a block away from a charming cobblestone area full of art galleries, antique stores, and ethnic restaurants. But this was all an ex*pense*, as she put it, with heavy emphasis on the second syllable, not just financial expense, but a too-extravagant experience of life. That evening, as I watched the sunset's pinwheels of apricot and mauve slowly explode into red ribbons, I thought: *The sensory misers will inherit the earth, but first they will make it not worth living on.*

When you consider something like death, after which (there being no news flash to the contrary) we may well go out like a candle flame, then it probably doesn't matter if we try too hard, are awkward sometimes, care for one another too deeply, are excessively curious about nature, are too open to experience, enjoy a nonstop expense of the senses in an effort to know life intimately and lovingly. It probably doesn't matter if, while trying to be modest and eager watchers of life's many spectacles, we sometimes look clumsy or get dirty or ask stupid questions or reveal our ignorance or say the wrong thing or light up with wonder like the children we all are. It probably doesn't matter if a passerby sees us dipping a finger into the moist pouches of dozens of lady's slippers to find out what bugs tend to fall into them, and thinks us a bit eccentric. Or a neighbor, fetching her mail, sees us standing in the cold with our own letters in one hand and a seismically red autumn leaf in the other, its color hitting our senses like a blow from a stun gun, as we stand with a huge grin, too paralyzed by the intricately veined gaudiness of the leaf to move.

WHY LEAVES TURN COLOR IN THE FALL

The stealth of autumn catches one unaware. Was that a goldfinch perching in the early September woods, or just the first turning leaf? A red-winged blackbird or a sugar maple closing up shop for the winter? Keen-eyed as leopards, we stand still and squint hard, looking for signs of movement. Early-morning frost sits heavily on the grass, and turns barbed wire into a string of stars. On a distant hill, a small square of yellow appears to be a lighted stage. At last the truth dawns on us: Fall is staggering in, right on schedule, with its baggage of chilly nights, macabre holidays, and spectacular, heart-stoppingly beautiful leaves. Soon the leaves will start cringing on the trees, and roll up in clenched fists before they actually fall off. Dry seedpods will rattle like tiny gourds. But first there will be weeks of gushing color so bright, so pastel, so confettilike, that people will travel up and down the East Coast just to stare at it—a whole season of leaves.

Where do the colors come from? Sunlight rules most living things with its golden edicts. When the days begin to shorten, soon after the summer solstice on June 21, a tree reconsiders its leaves. All summer it feeds them so they can process sunlight, but in the dog days of summer the tree begins pulling nutrients back into its trunk and roots, pares down, and gradually chokes off its leaves. A corky layer of cells forms at the leaves' slender petioles, then scars over. Undernourished, the leaves stop producing the pigment chlorophyll, and photosynthesis ceases. Animals can migrate, hibernate, or store food to prepare for winter. But where can a tree go? It survives by dropping its leaves, and by the end of autumn only a few fragile threads of fluid-carrying xylem hold leaves to their stems.

A turning leaf stays partly green at first, then reveals splotches of yellow and red as the chlorophyll gradually breaks down. Dark green seems to stay longest in the veins, outlining and defining them. During the summer, chlorophyll dissolves in the heat and light, but it is also being steadily replaced. In the fall, on the other hand, no new pigment is produced, and so we notice the other colors that were always there, right in the leaf, although chlorophyll's shocking green hid them from view. With their camouflage gone, we see these

258 ≈ Diane Ackerman

colors for the first time all year, and marvel, but they were always there, hidden like a vivid secret beneath the hot glowing greens of summer.

The most spectacular range of fall foliage occurs in the northeastern United States and in eastern China, where the leaves are robustly colored, thanks in part to a rich climate. European maples don't achieve the same flaming reds as their American relatives, which thrive on cold nights and sunny days. In Europe, the warm, humid weather turns the leaves brown or mildly yellow. Anthocyanin, the pigment that gives apples their red and turns leaves red or red-violet, is produced by sugars that remain in the leaf after the supply of nutrients dwindles. Unlike the carotenoids, which color carrots, squash, and corn, and turn leaves orange and yellow, anthocyanin varies from year to year, depending on the temperature and amount of sunlight. The fiercest colors occur in years when the fall sunlight is strongest and the nights are cool and dry (a state of grace scientists find vexing to forecast). This is also why leaves appear dizzyingly bright and clear on a sunny fall day: The anthocyanin flashes like a marquee.

Not all leaves turn the same colors. Elms, weeping willows, and the ancient ginkgo all grow radiant yellow, along with hickories, aspens, bottlebrush buckeyes, cottonweeds, and tall, keening poplars. Basswood turns bronze, birches bright gold. Water-loving maples put on a symphonic display of scarlets. Sumacs turn red, too, as do flowering dogwoods, black gums, and sweet gums. Though some oaks yellow, most turn a pinkish brown. The farmlands also change color, as tepees of cornstalks and bales of shredded-wheat-textured hay stand drying in the fields. In some spots, one slope of a hill may be green and the other already in bright color, because the hillside facing south gets more sun and heat than the northern one.

An odd feature of the colors is that they don't seem to have any special purpose. We are predisposed to respond to their beauty, of course. They shimmer with the colors of sunset, spring flowers, the tawny buff of a colt's pretty rump, the shuddering pink of a blush. Animals and flowers color for a reason—adaptation to their environment—but there is no adaptive reason for leaves to color so beautifully in the fall any more than there is for the sky or ocean to be blue. It's just one of the haphazard marvels the planet bestows every year. We find the sizzling colors thrilling, and in a sense they dupe us. Colored like living things, they signal death and disintegration. In time, they will become fragile and, like the body, return to dust. They are as we hope our own fate will be when we die: Not to vanish, just to sublime from one beautiful state into another. Though leaves lose their green life, they bloom with urgent colors, as the woods grow mummified day by day, and Nature becomes more carnal, mute, and radiant.

We call the season "fall," from the Old English *feallan,* to fall, which leads back through time to the Indo-European *phol,* which also means to fall. So the word and the idea are both extremely ancient, and haven't really changed since the first of our kind needed a name for fall's leafy abundance. As we say the word, we're reminded of that other Fall, in the garden of Eden, when fig leaves never withered and scales fell from our eyes. Fall is the time when leaves fall from the trees, just as spring is when flowers spring up, summer is when we simmer, and winter is when we whine from the cold.

Children love to play in piles of leaves, hurling them into the air like confetti, leaping into soft unruly mattresses of them. For children, leaf fall is just one of the odder figments of Nature, like hailstones or snowflakes. Walk down a lane overhung with trees in the never-never land of autumn, and you will forget about time and death, lost in the sheer delicious spill of color. Adam and Eve concealed their nakedness with leaves, remember? Leaves have always hidden our awkward secrets.

But how do the colored leaves fall? As a leaf ages, the growth hormone, auxin, fades, and cells at the base of the petiole divide. Two or three rows of small cells, lying at right angles to the axis of the petiole, react with water, then come apart, leaving the petioles hanging on by only a few threads of xylem. A light breeze, and the leaves are airborne. They glide and swoop, rocking in invisible cradles. They are all wing and may flutter from yard to yard on small whirlwinds or updrafts, swiveling as they go. Firmly tethered to earth, we love to see things rise up and fly—soap bubbles, balloons, birds, fall leaves. They remind us that the end of a season is capricious, as is the end of life. We especially like the way leaves rock, careen, and swoop as they fall. Everyone knows the motion. Pilots sometimes do a maneuver called a "falling leaf," in which the plane loses altitude quickly and on purpose, by slipping first to the right, then to the left. The machine weighs a ton or more, but in one pilot's mind it is a weightless thing, a falling leaf. She has seen the motion before, in the Vermont woods where she played as a child. Below her the trees radiate gold, copper, and red. Leaves are falling, although she can't see them fall, as she falls, swooping down for a closer view.

At last the leaves leave. But first they turn color and thrill us for weeks on end. Then they crunch and crackle underfoot. They *shush*, as children drag their small feet through leaves heaped along the curb. Dark, slimy mats of leaves cling to one's heels after a rain. A damp, stuccolike mortar of semidecayed leaves protects the tender shoots with a roof until spring, and makes a rich humus. An occasional bulge or ripple in the leafy mounds signals a shrew or a field mouse tunneling out of sight. Sometimes one finds in fossil stones the imprint of a leaf, long since disintegrated, whose outlines remind us how detailed, vibrant, and alive are the things of this earth that perish.

ANIMALS

Polar bears are not white, they're clear. Their transparent fur doesn't contain a white pigment, but the hair shafts house many tiny air bubbles, which scatter the sun's white light, and we register the spectacle as white fur. The same thing happens with a swan's white feathers, and the white wings of some butterflies. We tend to think of everything on earth as having its own deep-down rich color, but even razzmatazz colors that hit one's eyes like carefully aimed fireworks are just a thin rind on things, the merest layer of pigment. And many objects have no pigment at all, but seem richly colored nonetheless because of tricks played by our eyes. Just as the oceans and sky are blue because of the scattering of light rays, so are a blue jay's feathers, which contain no blue pigment. The same is true of the blue on a turkey's neck, the blue on the tail of the blue-tailed skink, the blue on a baboon's rump. Grass and leaves, on the other hand, are inherently green because of the green pigment chlorophyll. The tropical rain forests and the northern woods both sing a green anthem. Against a backdrop of chlorophyll green, earth brown, and sky-and-water blue, animals have evolved kaleidoscopic colors to attract mates, disguise themselves, warn off would-be predators, scare rivals away from their territory, signal a parent that it's time to be fed. Woodland birds are often drably colored and lightly speckled, to blend in with the branches and sifting sunlight. There are lots of "LBJs," or "little brown jobs," as birders sometimes call them.

Abbott Thayer, an early twentieth-century artist and naturalist, noticed what he called *countershading,* a natural camouflaging that makes animals most brightly colored on the parts of their body that are least exposed to sunlight, and darker on those areas that are most exposed. A good example is the penguin, which is white on the breast so that it will look like pale sky when viewed from underneath in the ocean, and black on its back, so that it will blend in with the dark depths of the ocean when viewed from on top. Since penguins are not in much danger from land predators, their obvious two-tone linoleum-floor look doesn't matter when they're waddling on shore. Camouflage and display is the name of the game in the animal kingdom. Insects are especially good at disguise; one famous example is the British peppered moth, which took only fifty years to change from a lackluster salt-and-pepper gray to nearly black so that it could blend in with tree bark that had become stained by industrial pollution. Pale moths were easier for a bird to spot as the tree trunks grew darker, and so darker moths survived to produce even darker moths, which in turn survived. Animals will do most anything to disguise themselves: Many fish have what look like eyes on their tails so that a predator will aim its attack on a less vital part of the body; some grasshoppers look so much like quartz they become invisible on

South African hills; clever butterflies sport large, dark eyespots on their wings, so that a songbird predator will think it's facing an owl; the insects called walking sticks appear dark and gnarly as twigs; Kenyan bush crickets blend in with the lichens on a tree trunk; katydids green up like leaves—some species even develop brown fungusy-looking sections; a Peruvian grasshopper mimics the crinkled dead leaves on the forest floor; the Malaysian tussock moth has wings that resemble decaying leaves: brown, torn, or perforated. Various insects costume themselves as snakes, others as bird droppings; lizards, shrimp, frogs, fish, and a few spiders tint their body color to blend in with their surroundings. Camouflage to a fish means scintillating like the water that surrounds it, breaking up the apparent outline of its body, and vanishing among the corridors of down-welling light. As Sandra Sinclair explains it in *How Animals See:* "Each scale reflects one-third of the spectrum; where three scales overlap, all colors are canceled out, leaving a mirrorlike effect." All a predator may see is a twisting flash of light. Luminescent squids maneuver at depths where there is little light; swimming through the gloom, they mimic the natural light from above, and can even disguise themselves as clouds floating over the surface of the water in order to become invisible to their prey. They are "stealth" squids. All sorts of animals can change color quickly by shrinking or enlarging their store of melanin; they either spread the color around so much that they look darker, or tug the color into a smaller space so that some underlying pigment becomes visible. In *Speak, Memory*, Vladimir Nabokov writes joyously of his fascination with the mimicry of moths and butterflies:

> Consider the imitation of oozing poison by bubblelike macules on a wing . . . or by glossy yellow knobs on a chrysalis ("Don't eat me—I have already been squashed, sampled and rejected"). Consider the tricks of an acrobatic caterpillar (of the Lobster Moth) which in infancy looks like bird's dung. . . . When a certain moth resembles a certain wasp in shape and color, it also walks and moves its antennae in a waspish, unmothlike manner. When a butterfly has to look like a leaf, not only are all the details of a leaf

beautifully rendered but markings mimicking grub-bored holes are generously thrown in. "Natural selection," in the Darwinian sense, could not explain the miraculous coincidence of imitative aspect and imitative behavior, nor could one appeal to the theory of the "struggle for life" when a protective device was carried to a point of mimetic subtlety, exuberance, and luxury far in excess of a predator's power of appreciation. I discovered in nature the nonutilitarian delights that I sought in art. Both were a form of magic, both were a game of intricate enchantment and deception.

Animals indulge in such lavish and luscious forms of display that it would take a whole book just to list their color-mad graces. The peacock's scintillating, many-eyed tail is so famous an example it's become eponymous. "What a peacock he is!" we say of a gentleman dandied up beyond belief. Color as a silent language works so well that nearly every animal speaks it. Octopuses change color as they change mood. A scared freshwater perch automatically turns pale. A king penguin chick knows to peck at the apricot comet on its parent's bill if it wants to be fed. A baboon flashes its blue rump in sexual or submissive situations. Confront a male robin with a handful of red feathers and it will attack it. A deer pops its white tail as a warning to its kin and then springs out of the yard. We lift our eyebrows to signal our disbelief. But many animals wear their gaudy colors as warnings, as well. The arrowpoison frog, which dwells in the Amazon rain forest, glistens with vibrant aqua blue and scarlet. *Don't mess with me!* its color shrieks at would-be predators. I was with a group of people who came upon such a frog squatting on a log, and the temptation to touch its cloisonné-like back was so strong one man automatically began to reach out for it when his neighbor grabbed his wrist, just in time. That frog didn't need to flee; it was coated with a slime so poisonous that if the man had touched it, and then touched his eye or mouth, he would have been poisoned on the spot.

When your cat stalks a low-lying slither at twilight, it's tempting to believe the old wives' tale that cats can see in the dark. After all, don't their eyes glow? But no animal can see without light. Cats, and other night-roving creatures, have a thin, iridescent* layer of reflecting cells behind the retina called the tapetum. Light strikes its mirror surface and bounces back at the retina, allowing an animal to see in faint light. If you hold a flashlight against your forehead at night and shine its light into the forest or along a swamp or ocean, you're bound to "shine" the red or amber eyes of some nocturnal creature—a spider, a caiman, a cat, a moth, a bird. Even scallops, with their tiny stuffed-olive-looking eyes, have a tapetum to capture more light, so that late at night they can observe any whelk sneaking up on them. Results of scientific experiments seem to indicate that cold-blooded animals can see better in dim light than warm-blooded ones, so amphibians generally have better night vision than mammals. (In one test conducted by researchers from the University of Copenhagen and the University of Helsinki, humans needed eight times as much light to see a worm at night than a toad did.) Cats, like other predators, have their eyes set squarely in front; they often have relatively big eyes and great depth perception, so that they can sight and track their prey. Consider the owl, a pair of binoculars with wings, whose eyes make up a third of its head size. Arrowhead crabs, bright spiderlike reef creatures familiar to scuba-divers, have eyes set so far apart they can see in almost a complete circle. Horses have little depth perception, because their eyes are placed far around each side of the head. Like prey in general, they need peripheral vision to keep an eye out for an attack from a predator. I've always thought it was particularly brave of horses to be willing to take jumps they must lose sight of at the last moment. Predators frequently have vertical pupils, since they look forward for their prey; whereas sheep, goats, and many other hoofed animals, which must be vigilant across the fields in which they graze, have horizontal pupils. An interesting feature of the alligator's pupil is that it can tilt a little as the angle of the head changes, so that prey will always be in focus. Roadside alligator wrestlers who flip a 'gator over, rub its stomach, and "put it to sleep" are actually giving it a bad case of vertigo. Upside down,

---

*From Latin *iris*, rainbow + *escence*, becoming. The combination *-esc-* converts words from a static state to one of motion and process: putrescence, adolescence, luminescence.

an alligator's pupils can't adjust, and the world becomes a confusing tumult of images. Many insects have compound eyes that iridesce, but few are as beautiful as the eye of the goldeneye lacewing: a background of black topped by a perfect six-pointed star, which shimmers blue at its tips, green as you move inward, then yellow, and finally red at the center.

Prairie dogs are color-blind to red and green, owls are entirely colorblind (because they have only rod cells), and ants don't see red at all. The deer that stroll into my yard to feast on apples and rosebushes see me mainly as shades of gray, as do the rabbits that eat the wild strawberries on my back lawn and are tame enough to kick in the rump. A surprising number of animals do see in color, but the colors they see are different. Unlike us, some also see in infrared, or with radically different kinds of eyes (barred, compound, iridescent, tubular, at the ends of stalks). The world that greets them looks different. Horror films persuaded us that the fly's compound eye meant that it saw the same image repeated many times, but scientists have now taken pictures through the eyes of insects, and we know that a fly sees a single complete scene, as we do, only a greatly curved one: It would be the equivalent of looking at the world through a glass paperweight. We assume that insects and animals don't see very well, but birds can see the stars, some butterflies can see in the ultraviolet range, and some jellyfish create their own light to read by. Bees can judge the angle at which light hits their photoreceptors, and therefore locate the position of the sun in the sky, even on a partly cloudy day. There are orchids that look so much like bees that bees try to mate with them, spreading pollen in the process. This intricate and extreme adaptation wouldn't work if bees' vision were poor. The reason movies appear to be continuous is that they move at about twenty-four frames per second, whereas we process images at fifty to sixty per second. When we watch a movie, we're actually watching a blank screen for about half the time. The rest of the time, many still photographs are flashed one after another, each slightly different and yet related to the preceding one. The eye dawdles over each photograph just long enough to slur into the next one, and they seem to be a single continuously moving picture. The eye persists in linking up the separate images. Bees, on the other hand, are used to images flashing at three hundred per second, so *Lawrence of Arabia* would be just a series of stills to them. It used to be thought that a bee's "waggle dance" included semaphore instructions for how to get to the great feeding places the bee had just been to; but now scientists think that the waggle dance also conveys messages in touch, smell, and hearing. Although it's true that bees can see in ultraviolet, they're weak on the red end of the spectrum, so a white flower looks blue to a bee, and a red flower is of little interest. Moths, birds, and bats, on the other hand, adore red flowers. Flowers that look drab and simple to us—nothing but white petals—to a bee may be lit up like a billboard flanked by neon signs pointing the way to the nectar. Bulls don't have color vision, so the bright red of the matador's cape could just as easily be black or orange. Red is for the benefit of the human audience, which finds the color intrinsically arousing and also suggestive of the soon-to-be-flowing blood of either the bull or the matador. The bull just focuses irritably on the large object moving in front of the man and charges.

The Boran people of Kenya are led to honeybees' nests by the pantomiming of a bird, the African honey guide *(Indicator indicator).* If the Boran are in the mood for honey, they whistle to call the bird. Or, if the bird is hungry for honey, it flies around the Boran, alerting them with its *"tirr-tirr-tirr."* Then it disappears briefly, apparently to check on the whereabouts of a honeybee nest, and returns to guide them with short flights and repeated calls. When the bird gets to the nest, it flies down to indicate the right spot and changes its call. Skillfully, the Boran break into the nest and take honey; they leave plenty for the bird, which would otherwise find the nest hard to invade. German ornithologists at the Max-Planck-Institut, who spent three years studying this strange symbiotic relationship, discovered that it takes the tribesmen almost three times as long to find honey without the help of the honey birds. Apparently, the birds also guide honey badgers in a similar way. Animals' eyes may be quick and keen, but few eyes are as probing as those of the artist, another species of hunter, whose prey lives in both the outer world and the inner tundra.

THE PAINTER'S EYE

In his later years, Cézanne suffered a famous paroxysm of doubt about his genius. Could his art have been only an eccentricity of his vision, not imagination and talent guarded by a vigilant esthetic? In his excellent essay on Cézanne in *Sense and Nonsense,* Maurice Merleau-Ponty says: "As he grew old, he wondered whether the novelty of his painting might not come from trouble with his eyes, whether his whole life had not been based upon an accident of the body." Cézanne anxiously considered each brush stroke, striving for the fullest sense of the world, as Merleau-Ponty describes so well:

> We *see* the depth, the smoothness, the softness, the hardness of objects; Cézanne even claimed that we see their odor. If the painter is to express the world, the arrangement of his colors must carry with it this invisible whole, or else his picture will only hint at things and will not give them in the imperious unity, the presence, the insurpassable plenitude which is for us the definition of the real. That is why each brush stroke must satisfy an infinite number of conditions. Cézanne sometimes pondered for hours at a time before putting down a certain stroke, for, as Bernard said, each stroke must "contain the air, the light, the object, the composition, the character, the outline, and the style." Expressing what *exists* is an endless task.

Opening up wide to the fullness of life, Cézanne felt himself to be the conduit where nature and humanity met—"The landscape thinks itself in me . . . I am its consciousness"—and would work on all the different sections of a painting at the same time, as if in that way he could capture the many angles, half-truths, and reflections a scene held, and fuse them into one conglomerate version. "He considered himself powerless," Merleau-Ponty writes, "because he was not omnipotent, because he was not God and wanted neverthe-less to portray the world, to change it completely into a spectacle, to make *visible* how the world *touches* us." When one thinks of the masses of color and shape in his paintings, perhaps it won't come as a surprise to learn that Cézanne was myopic, although he refused glasses, reputedly crying "Take those vulgar things away!" He also suffered from diabetes, which may have resulted in some retinal damage, and in time he developed cataracts (a clouding of the clear lens). Huysmans once captiously described him as "An artist with a diseased retina, who, exasperated by a defective vision, discovered the basis of a new art." Born into a different universe than most people, Cézanne painted the world his slightly askew eyes saw, but the random chance of that possibility gnawed at him. The sculptor Giacometti, on the other hand, whose long, stretched-out figures look as consciously distorted as one could wish, once confessed amiably: "All the critics spoke about the metaphysical content or the poetic message of my work. But for me it is nothing of the sort. It is a purely optical exercise. I try to represent a head as I see it."

Quite a lot has been learned in recent years about the vision problems of certain artists, whose eyeglasses and medical records have survived. Van Gogh's "Irises" sold at Christie's in 1988 for forty-nine million dollars, which would surely have amused him, since he sold only one painting during his lifetime. Though he was known for cutting off his ear, van Gogh also hit himself with a club, went to many church services each Sunday, slept on a board, had bizarre religious hallucinations, drank kerosene, and ate paint. Some researchers now feel that a few of van Gogh's stylistic quirks (coronas around streetlamps, for instance) may not have been intentional distortions at all but the result of illness, or, indeed, of poisoning from the paint thinners and resins he used, which could have damaged his eyes so that he saw halo effects around light sources. According to Patrick Trevor-Roper, whose *The World Through Blunted Sight* investigates the vision problems of painters and poets, some of the possible diagnoses for van Gogh's depression "have included cerebral tumour, syphilis, magnesium deficiency, temporal lobe epilepsy, poisoning by digitalis (given as a treatment for epilepsy, which could have provoked the yellow vision), and glaucoma (some self-portraits show a dilated right pupil, and he depicted coloured haloes around lights)." Most recently, a scientist speaking before a meeting of neurologists in Boston added Geschwind's syndrome, a personality disorder that sometimes accompanies epilepsy. Van Gogh's own doctor said of him: "Genius and lunacy are well known next-door neighbors." Many of those ailments could have affected his vision. But, equally important, the most brilliant pigments used to include toxic heavy metals like copper, cadmium, and mercury. Fumes and poisons could easily get into food, since painters frequently worked and lived in the same rooms. When the eighteenth-century animal painter George Stubbs went on his honeymoon, he stayed in a two-room cottage, in one room of which he hung up the decaying carcass of a horse, which in free moments he studiously dissected. Renoir was a heavy smoker, and he probably didn't bother to wash his hands before he rolled a cigarette; paint from his fingers undoubtedly rubbed onto the paper. Two Danish internists, studying the relationship between arthritis and heavy metals, have compared the color choices in paintings by Renoir, Peter Paul Rubens, and Raoul Dufy (all rheumatoid arthritis sufferers), with those of their contemporaries. When Renoir chose his bright reds, oranges, and blues, he was also choosing big doses of aluminum, mercury, and cobalt. In fact, up to 60 percent of the colors Renoir preferred contained dangerous metals, twice the amount used by such contemporaries of his as Claude Monet or Edgar Degas, who often painted with darker pigments made from safer iron compounds.

According to Trevor-Roper, there is a myopic personality that artists, mathematicians, and bookish people tend to share. They have "an interior life different from others," a different personality, because only the close-up world is visually available to them. The imagery in their work tends to pivot around things that "can be viewed at very close range," and they're more introverted. Of Degas's myopia, for example, he says:

> As time passed he was often reduced to painting in pastel rather than oil as being an easier medium for his failing sight. Later, he discovered that by using photographs of the models or horses he sought to depict, he was able to bring these comfortably within his limited focal range. And finally he fell back increasingly on sculpture where at least he could be sure that his sense of touch would always remain true, saying, 'I must learn a blind man's trade now,' although he had always in fact had an interest in modelling.

Trevor-Roper points out that the mechanism which causes short-sightedness (an elongated eye) affects perception of color as well (reds will appear more starkly defined); cataracts, especially, may affect color, blurring and reddening simultaneously. Consider Turner, whose later paintings Mark Twain once described as "like a ginger cat having a fit in a bowl of tomatoes." Or Renoir's "increasing fascination for reds." Or Monet, who developed such severe cataracts that he had to label his tubes of paint and arrange colors carefully on his palette. After a cataract operation, Monet is reported by friends to have been surprised by all the blueness in the world, and to have been appalled by the strange colors in his recent work, which he anxiously retouched.

One theory about artistic creation is that extraordinary artists come into this world with a different way of seeing. That doesn't explain genius, of course, which has so much to do with risk, anger, a blazing emotional furnace, a sense of esthetic decorum, a savage wistfulness, lidless curiosity, and many other qualities, including a willingness to be fully available to life, to pause over both its general patterns and its ravishing details. As the robustly sensuous painter Georgia O'Keeffe once said: "In a way, nobody sees a flower really, it is so small, we haven't time—and to see takes time, like to have a friend takes time." What kind of novel vision do artists bring into the world with them, long before they develop an inner vision? That question disturbed Cézanne, as it has other artists—as if it made any difference to how and what he would end up painting. When all is said and done, it's as Merleau-Ponty says: *"This work to be done called for this life."*

THE FACE OF BEAUTY

In a study in which men were asked to look at photographs of pretty women, it was found they greatly preferred pictures of women whose pupils were dilated. Such pictures caused the pupils of the men's eyes to dilate as much as 30 percent. Of course, this is old news to women of the Italian Renaissance and Victorian England alike, who used to drop belladonna (a poisonous plant in the nightshade family, whose name means "beautiful woman") into their eyes to enlarge their pupils before they went out with gentlemen. Our pupils expand involuntarily when we're aroused or excited; thus, just seeing a pretty woman with dilated pupils signaled the men that she found them attractive, and that made their pupils begin a body-language tango in reply. When I was on shipboard recently, traveling through the ferocious winds and waves of Drake Passage and the sometimes bouncy waters around the Antarctic peninsula, the South Orkneys, South Georgia, and the Falklands, I noticed that many passengers wore a scopolamine patch behind one ear to combat seasickness. Greatly dilated pupils, a side effect of the patch, began to appear a few days into the trip; everybody one met had large, welcoming eyes, which no doubt encouraged the feeling of immediate friendship and camaraderie. Some people grew to look quite zombielike, as they drank in wide gulps of light, but most seemed especially open and warm.* Had they checked, the women would have discovered that their cervixes were dilated, too. In professions where emotion or sincere interests need to be hidden, such as gambling or jade-dealing, people often wear dark glasses to hide intentions visible in their telltale pupils.

We may pretend that beauty is only skin deep, but Aristotle was right when he observed that "beauty is a far greater recommendation than any letter of introduction." The sad truth is that attractive people do better in school, where they receive more help, better grades, and less punishment; at work, where they are rewarded with higher pay, more prestigious jobs, and faster promotions; in finding mates, where they tend to be in control of the relationships and make most of the decisions; and among total strangers, who assume them to be interesting, honest, virtuous, and successful. After all, in fairy tales, the first stories most of us hear, the heroes are handsome, the heroines are beautiful, and the wicked sots are ugly. Children learn implicitly that good people are beautiful and bad people are

---

*An alkaloid extracted from henbane and various other plants of the nightshade family, scopolamine has also been used as truth serum. What a perfect cocktail for a cruise: large pupils continuously signaling interest in everyone they see, and a strong urge to be uninhibited and open to persuasion.

ugly, and society restates that message in many subtle ways as they grow older. So perhaps it's not surprising that handsome cadets at West Point achieve a higher rank by the time they graduate, or that a judge is more likely to give an attractive criminal a shorter sentence. In a 1968 study conducted in the New York City prison system, men with scars, deformities, and other physical defects were divided into three groups. The first group received cosmetic surgery, the second intensive counseling and therapy, and the third no treatment at all. A year later, when the researchers checked to see how the men were doing, they discovered that those who had received cosmetic surgery had adjusted the best and were less likely to return to prison. In experiments conducted by corporations, when different photos were attached to the same résumé, the more attractive person was hired. Prettier babies are treated better than homelier ones, not just by strangers but by the baby's parents as well. Mothers snuggle, kiss, talk to, play more with their baby if it's cute; and fathers of cute babies are also more involved with them. Attractive children get higher grades on their achievement tests, probably because their good looks win praise, attention, and encouragement from adults. In a 1975 study, teachers were asked to evaluate the records of an eight-year-old who had a low IQ and poor grades. Every teacher saw the same records, but to some the photo of a pretty child was attached, and to others that of a homely one. The teachers were more likely to recommend that the homely child be sent to a class for retarded children. The beauty of another can be a valuable accessory. One particularly interesting study asked people to look at a photo of a man and a woman, and to evaluate only the man. As it turned out, if the woman on the man's arm was pretty, the man was thought to be more intelligent and successful than if the woman was unattractive.

Shocking as the results of these and similar experiments might be, they confirm what we've known for ages: Like it or not, a woman's face has always been to some extent a commodity. A beautiful woman is often able to marry her way out of a lower class and poverty. We remember legendary beauties like Cleopatra and Helen of Troy as symbols of how beauty can be powerful enough to cause the downfall of great leaders and change the career of empires. American women spend millions on makeup each year; in addition, there are the hairdressers, the exercise classes, the diets, the clothes. Handsome men do better as well, but for a man the real commodity is height. One study followed the professional lives of 17,000 men. Those who were at least six feet tall did much better—received more money, were promoted faster, rose to more prestigious positions. Perhaps tall men trigger childhood memories of looking up to authority—only our parents and other adults were tall, and they had all the power to punish or protect, to give absolute love, set our wishes in motion, or block our hopes.

The human ideal of a pretty face varies from culture to culture, of course, and over time, as Abraham Cowley noted in the seventeenth century:

> *Beauty, thou wild fantastic ape*
> *Who dost in every country change thy shape!*

But in general what we are probably looking for is a combination of mature and immature looks—the big eyes of a child, which make us feel protective, the high cheekbones and other features of a fully developed woman or man, which make us feel sexy. In an effort to look sexy, we pierce our noses, elongate our earlobes or necks, tattoo our skin, bind our feet, corset our ribs, dye our hair, have the fat liposuctioned from our thighs, and alter our bodies in countless other ways. Throughout most of western history, women were expected to be curvy, soft, and voluptuous, real earth mothers radiant with sensuous fertility. It was a preference with a strong evolutionary basis: A plump woman had a greater store of body fat and the nutrients needed for pregnancy, was more likely to survive during times of hunger, and would be able to protect her growing fetus and breastfeed it once it was born. In many areas of Africa and India, fat is considered not only beautiful but prestigious for both men and women. In the United States, in the Roaring Twenties and also in the Soaring Seventies and Eighties, when ultrathin was in, men wanted women to have the figures of teenage boys, and much psychological hay could be made from how this reflected the changing role of women in society and the work place. These days, most men I know prefer women to have a curvier, reasonably fit body, although most women I know would still prefer to be "too" thin.

But the face has always attracted an admirer's first glances, especially the eyes, which can be so smoldery and eloquent, and throughout the ages people have emphasized their facial features with makeup. Archaeologists have found evidence of Egyptian perfumeries and beauty parlors dating to 4,000 B.C., and makeup paraphernalia going back to 6,000 B.C. The ancient Egyptians preferred green eye shadow topped with a glitter made from crushing the iridescent carapaces of certain beetles; kohl eye liner and mascara; blue-black lipstick; red rouge; and fingers and feet stained with henna. They shaved their eyebrows and drew in false ones. A fashionable Egyptian woman of those days outlined the veins on her breasts in blue and coated her nipples with gold. Her nail polish signaled social status, red indicating the highest. Men also indulged in elaborate potions and beautifiers; and not only for a night out: Tutankhamen's tomb included jars of makeup and beauty creams for his use in the afterlife. Roman men adored cosmetics, and commanders had their hair coiffed and perfumed and their nails lacquered before they went into battle. Cosmetics appealed even more to Roman women, to one of whom Martial wrote in the first century A.D., "While you remain at home, Galla, your hair is at the hairdresser's; you take out your teeth at night and sleep tucked away in a hundred cosmetic boxes— even your face does not sleep with you. Then you wink at men under an eyebrow you took out of a drawer that same morning." A second-century Roman physician invented cold cream, the formula for which has changed little since then. We may remember from the Old Testament that Queen Jezebel painted her face before embarking on her wicked ways, a fashion she learned from the high-toned Phoenicians in about 850 B.C. In the eighteenth century, European women were willing to eat Arsenic Complexion Wafers to make their skin whiter; it poisoned the hemoglobin in the blood so that they developed a fragile, lunar whiteness. Rouges often contained such dangerous metals as lead and mercury, and when used as lip-stain they went straight into the bloodstream. Seventeenth-century European women and men sometimes wore beauty patches in the shape of hearts, suns, moons, and stars, applying them to their breasts and face, to draw an admirer's eye away from any imperfections, which, in that era, too often included smallpox scars.

Studies conducted recently at the University of Louisville asked college men what they considered to be the ideal components in a woman's face, and fed the results into a computer. They discovered that their ideal woman had wide cheekbones; eyes set high and wide apart; a smallish nose; high eyebrows; a small neat chin; and a smile that could fill half of the face. On faces deemed "pretty," each eye was one-fourteenth as high as the face, and three-tenths its width; the nose didn't occupy more than five percent of the face; the distance from the bottom lip to the chin was one fifth the height of the face, and the distance from the middle of the eye to the eyebrow was one-tenth the height of the face. Superimpose the faces of many beautiful women onto these computer ratios, and none will match up. What this geometry of beauty boils down to is a portrait of an ideal mother—a young, healthy woman. A mother had to be fertile, healthy, and energetic to protect her young and continue to bear lots of children, many of whom might die in infancy. Men drawn to such women had a stronger chance of their genes surviving. Capitalizing on the continuing subleties of that appeal, plastic surgeons sometimes advertise with extraordinary bluntness. A California surgeon, Dr. Vincent Forshan, once ran an eight-page color ad in *Los Angeles* magazine showing a gorgeous young woman with a large, high bosom, flat stomach, high, tight buttocks, and long sleek legs posing beside a red Ferrari. The headline over the photo ran: "Automobile by Ferrari . . . *body by Forshan.*" Question: What do those of us who aren't tall, flawlessly sculpted adolescents do? Answer: Console ourselves with how relative beauty can be. Although it wins our first praise and the helpless gift of our attention, it can curdle before our eyes in a matter of moments. I remember seeing Omar Sharif in *Doctor Zhivago* and *Lawrence of Arabia,* and thinking him astoundingly handsome. When I saw him being interviewed on television some months later, and heard him declare that his only interest in life was playing bridge, which is how he spent most of his spare time, to my great amazement he was transformed before my eyes into an unappealing man. Suddenly his eyes seemed rheumy and his chin stuck out too much and none of the pieces of his anatomy fell together in the right proportions. I've watched this alchemy work in reverse, too, when a not-particularly-attractive stranger opened his mouth to speak and became ravishing. Thank heavens for the arousing qualities of zest, intelligence, wit, curiosity, sweetness, passion, talent, and grace. Thank heavens that, though good looks may rally one's attention, a lasting sense of a person's beauty reveals itself in stages. Thank heavens, as Shakespeare puts it in *A Midsummer Night's Dream:* "Love looks not with the eyes, but with the mind."

We are not just lovers of one another's features, of course, but also of nature's. Our passion for beautiful flowers we owe entirely to insects, bats, and birds, since these pollinators and flowers evolved together; flowers use color to attract birds and insects that will pollinate them. We may breed flowers to the pitch of sense-pounding color and smell we prefer, and we've greatly changed the look of nature by doing so, but there is a special gloriousness we find only in nature at its most wild and untampered with. In our "sweet spontaneous earth," as e. e. cummings calls it, we find startling and intimate beauties that fill us with ecstasy. Perhaps, like him, we

> *notice the convulsed orange inch of moon*
> *perching on this silver minute of evening*

and our pulse suddenly charges like cavalry, or our eyes close in pleasure and, in a waking faint, we sigh before we know what's happening. The scene is so beautiful it deflates us. Moonlight can reassure us that there will be light enough to find our way over dark plains, or to escape a night-prowling beast. Sunset's fiery glow reminds us of the warmth in which we thrive. The gushing colors of flowers signal springtime and summer, when food is plentiful and all life is radiantly fertile. Brightly colored birds turn us on, sympathetically, with their sexual flash and dazzle, because we're atavists at heart and any sex pantomime reminds us of our own. Still, the essence of natural beauty is novelty and surprise. In cummings's poem, it is an unexpected "convulsed orange inch of moon" that awakens one's notice. When this happens, our sense of community widens—we belong not just to one another but to other species, other forms of matter. "That we find a crystal or a poppy beautiful means that we are less alone," John Berger writes in *The Sense of Sight,* "that we are more deeply inserted into existence than the course of a single life would lead us to believe." Naturalists often say that they never tire of seeing the same mile of rain forest, or of strolling along the same paths through the savanna. But, if you press them, they inevitably add that there is always something new to behold, that it is always different. As Berger puts it: "beauty is always an exception, always *in despite of.* This is why it moves us." And yet we also respond passionately to the highly organized way of behold-ing life we call art. To some extent Art is like trapping nature inside a paperweight. Suddenly a locale, or an abstract emotion, is viewable at one's leisure, falls out of flux, can be rotated and considered from different vantage points, becomes as fixed and to that extent as holy as the landscape. As Berger puts it:

> All the languages of art have been developed as an attempt to transform the instantaneous into the permanent. Art supposes that beauty is not an exception—is not *in despite of*—but is the basis for an order. . . . Art is an organized response to what nature allows us to glimpse occasionally. . . . the transcendental face of art is always a form of prayer.

Art is more complex than that, of course. Intense emotion is stress-ful, and we look to artists to feel for us, to suffer and rejoice, to describe the heights of their passionate response to life so that we can enjoy them from a safe distance, and get to know better what the full range of human experience really is. We may not choose to live out the extremes of consciousness we find in Jean Genet or Edvard Munch, but it's wonderful to peer into them. We look to artists to stop time for us, to break the cycle of birth and death and temporarily put an end to life's processes. It is too much of a whelm for any one person to face up to without going into sensory overload. Artists, on the other hand, court that intensity. We ask artists to fill our lives with a cavalcade of fresh sights and insights, the way life was for us when we were children and everything was new.* In time, much of life's spectacle becomes a polite blur, because if we stop to consider every speckle-throated lily we will never get our letters filed or pomegranates bought.

Unbeautiful things often delight our eyes, too. Gargoyles, glitz, intense slabs of color, organized tricks of light. Sparklers and fireworks are almost painful to watch, but we call them beautiful. A flawless seven-carat marquise diamond is pure scintillation, which we also call beautiful. Throughout history, people have crafted nature's rudest rocks into exquisite jewelry, obsessed with the way in which light penetrates a crystal. We may find diamonds and other gems visually magnificent, but seeing them the way we do is a recent innovation. It was only in the eighteenth century that the newly improved art of gem-cutting produced the glittery stones full of fire and dazzle we admire. Before that, even the crown jewels appeared dull and listless. But in the eighteenth century faceted cuts became fashionable, along with plunging necklines. In fact, women often wore jewels pinned to the necklines of their gowns so that each might draw attention to the other. Why should a gem strike us as beautiful? A diamond acts like a bunched prism. Light entering a diamond ricochets around inside it, reflects from the back of it, and spreads out its colors more ebulliently than through an ordinary glass prism. A skilled diamond cutter enables light to streak along inside the stone's many facets and shoot out of the jewel at angles. Turn the diamond in your hand, and you see one pure color followed by another. Variety is the pledge that matter makes to living things. We find life's energy, motion, and changing colors trapped in the small, dead space of a diamond, which one moment glitters like neon

---

*As Laurens van der Post observed among the Bushmen of the Kalahari, "I saw the reason why poetry, music and the arts are matters of survival—of life and death to all of us. . . . The arts are both guardians and makers of this chain; they are charged with maintaining the aboriginal movements in the latest edition of man; they make young and immediate what is first and oldest in the spirit of man."

and the next spews out sabers of light. Our sense of wonder ignites, things are in the wrong place, a magical bonfire has been lit, the nonliving comes to life in an unexpected flash and begins a small, brief dance among the flames. Watching faces or fireworks or a spaceship launch, the dance is slower, but the colors and lights grow achingly intense as they surround and upstage us in a fantasia of pure visual ecstasy.

WATCHING A NIGHT LAUNCH OF THE SPACE SHUTTLE

A huge glittering tower sparkles across the Florida marshlands. Floodlights reach into the heavens all around it, rolling out carpets of light. Helicopters and jets blink around the launch pad like insects drawn to flame. Oz never filled the sky with such diamond-studded improbability. Inside the cascading lights, a giant trellis holds a slender rocket to its heart, on each side a tall thermos bottle filled with solid fuel the color and feel of a hard eraser, and on its back a sharp-nosed space shuttle, clinging like the young of some exotic mammal. A full moon bulges low in the sky, its face turned toward the launch pad, its mouth open.

On the sober consoles of launch control, numbers count backward toward zero. When numbers vanish, and reverse time ends, something will disappear. Not the shuttle—that will stay with us through eyesight and radar, and be on the minds of dozens of tracking dishes worldwide, rolling their heads as if to relieve the anguish. For hours we have been standing on these Floridian bogs, longing for the blazing rapture of the moment ahead, longing to be jettisoned free from routine, and lifted, like the obelisk we launch, that much nearer the infinite. On the fog-wreathed banks of the Banana River, and by the roadside lookouts, we are waiting: 55,000 people are expected at the Space Center alone.

When floodlights die on the launch pad, camera shutters and mental shutters all open in the same instant. The air feels loose and damp. A hundred thousand eyes rush to one spot, where a glint below the booster rocket flares into a pinwheel of fire, a sparkler held by hand on the Fourth of July. White clouds shoot out in all directions, in a dust storm of flame, a gritty, swirling Sahara, burning from gray-white to an incandescent platinum so raw it makes your eyes squint, to a radiant gold so narcotic you forget how to blink. The air is full of bee stings, prickly and electric. Your pores start to itch. Hair stands up stiff on the back of your neck. It used to be that the launch pad would melt at lift-off, but now 300,000 gallons of water crash from aloft, burst from below. Steam clouds scent the air with a mineral ash. Crazed by reflection, the waterways turn the color of pounded brass. Thick cumulus clouds shimmy and build at ground level, where you don't expect to see thunderheads.

Seconds into the launch, an apricot *whoosh* pours out in spasms, like the rippling quarters of a palomino, and now outbleaches the sun, as clouds rise and pile like a Creation scene. Birds leap into the air along with moths and dragonflies and gnats and other winged creatures, all driven to panic by the clamor: booming, crackling, howling downwind. What is flight, that it can take place in the fragile wings of a moth, whose power station is a heart small as a computer chip? What is flight, that it can groan upward through 4.5 million pounds of dead weight on a colossal gantry? Close your eyes, and you hear the deafening *rat-a-tat-tat* of firecrackers, feel them arcing against your chest. Open your eyes, and you see a huge steel muscle dripping fire, as seven million pounds of thrust pauses a moment on a silver haunch, and then the bedlam clouds let rip. Iron struts blow over the launch pad like newspapers, and shock waves roll out, pounding their giant fists, pounding the marshes where birds shriek and fly, pounding against your chest, where a heart already rapid begins running clean away from you. The air feels tight as a drum, the molecules bouncing. Suddenly the space shuttle leaps high over the marshlands, away from the now frantic laughter of the loons, away from the reedy delirium of the insects and the open-mouthed awe of the spectators, many of whom are crying, as it rises on a waterfall of flame 700 feet long, shooting colossal sparks as it climbs in a golden halo that burns deep into memory.

Only ten minutes from lift-off, it will leave the security blanket of our atmosphere, and enter an orbit 184 miles up. This is not miraculous. After all, we humans began in an early tantrum of the universe, when our chemical makeup first took form. We evolved through accidents, happenstance, near misses, and good luck. We developed language, forged cities, mustered nations. Now we change the course of rivers and move mountains; we hold back trillions of tons of water with cement dams. We break into human chests and heads; operate on beating hearts and thinking brains. What is defying gravity compared to that? In orbit, there will be no night and day, no up and down. No one will have their "feet on the ground." No joke will be "earthy." No point will be "timely." No thrill will be "out of this world." In orbit, the sun will rise every hour and a half, and there will be 112 days to each week. But then time has always been one of our boldest and most ingenious inventions, and, when you think about it, one of the least plausible of our fictions.

Lunging to the east out over the water, the shuttle rolls slowly onto its back, climbing at three g's, an upshooting torch, twisting an umbilical of white cloud beneath it. When the two solid rockets fall free, they hover to one side like bright red quotation marks, beginning an utterance it will take four days to finish. For over six minutes of seismic wonder it is still visible, this star we hurl up at the star-studded sky. What is a neighborhood? one wonders. Is it the clump of wild daisies beside the Banana River, in which moths hover and dive without the aid of rockets? For large minds, the Earth is a small place. Not small enough to exhaust in one lifetime, but a compact home, cozy, buoyant, a place to cherish, the spectral center of our life. But how could we stay at home forever?

THE FORCE OF AN IMAGE: RING CYCLE

In our mind's eye, that abstract seat of imagining, we picture the face of a lover, savor a kiss. When we think about him in passing, we have various thoughts; but when we actually picture him, as if he were a hologram, we feel a flush of emotion. There is much more to seeing than mere seeing. The visual image is a kind of tripwire for the emotions. One photo can remind us of an entire political regime, a war, a heroic moment, a tragedy. One gesture can symbol-ize the wide angles of parental love, the uncertainty and disorder of romantic love, the fun-house mirrors of adolescence, the quick trans-fusion of hope, the feeling of low-level wind shear in the heart we call loss. Look at a grassy hillside, and you can remember immedi-ately what freshly cut grass smells like, how it feels when it's damp, the stains it leaves on your jeans, the sound you can make blowing over a grass blade held just so between your thumbs, and other assorted memories associated with grass: picnicking with the family; playing dodge ball in an orchard in the Midwest; herding cattle from the dusty New Mexico desert up to high fields of lush green to graze; hiking through the Adirondacks; making love in a grassy field at the top of a hill, on a hot, breezy summer day, when the sun, shining through the clouds, lights one part of the hillside at a time, as if it were a room in which the lamp had been turned on. When we see an object, the whole peninsula of our senses wakes up to appraise the new sight. All the brain's shopkeepers consider it from their point of view, all the civil servants, all the accountants, all the students, all the farmers, all the mechanics. Together they all see the same sight—a grassy hillside—and each does a slightly different take on it, all of which adds up to what we see. Our other senses can trigger memories and emotions, too, but the eyes are especially good at symbolic, aphoristic, many-faceted perceiving. Knowing this, gov-ernments are forever erecting monuments. Generally they don't look like much, but people stand in front of them and rush with emotion anyway. The eye regards most of life as monumental. And some shapes affect us much more than others.

For example, I've been following the space program closely for the past twenty years, and learning with robust delight about the solar system, thanks mainly to the Voyager spacecraft, which have been sending back home movies of Earth's closest relatives. What a lovely shock it's been to discover that half the planets have rings: not just Saturn, but Jupiter, Uranus, Neptune, and maybe even Pluto. And all the rings are different. Jupiter's dark, narrow rings contrast with Saturn's bright broad ribbons. Uranus's obsidian rings have baguette moons in tow. The solar system has quietly been running rings around all of us. How magical and how poignant. Few symbols have ever meant as much to us, regardless of our religion, politics, age, or gender, as rings. We give rings to symbolize infinite love and the close harmony of two souls. Rings remind us of the simple cells that were the oldest version of life, and the symphony of cells we now are. We reach for the rings on merry-go-rounds. Rings halo what is sacred. We draw rings around things to emphasize them. Sports often take place in the magic ring of the playing field. A sensory kaleidoscope unfolds in the circus ring. Rings symbolize the infinite: We are only ever beginning to end. Rings signal a pledge made, a vow taken. Rings suggest eternity, agelessness, and perfection. We chart time on the face of a clock, as points along a ring. On playgrounds, children shoot marbles into a chalked circle; they are prime movers, acting out planetary mechanics. We bring the world into focus with the globes of our eyes, worlds within worlds. We treasure the well-rounded soul we think we see in a loved one. We believe that, just as a strong circle can be made out of two weaker arcs, we can complete ourselves by linking our life to someone else's. We who crave the no-loose-ends, deathless symmetry of a ring praise the wonders of the universe as best we can, traveling along the ring of birth and death. The Apollo astronauts returned to earth changed by seeing the home planet floating in space. What they saw was a kind of visual aphorism, and it's one we all need to learn by heart.

THE ROUND WALLS OF HOME

Picture this: Everyone you've ever known, everyone you've ever loved, your whole experience of life floating in one place, on a single planet underneath you. On that dazzling oasis, swirling with blues and whites, the weather systems form and travel. You watch the clouds tingle and swell above the Amazon, and know the weather that develops there will affect the crop yield half a planet away in Russia and China. Volcanic eruptions make tiny spangles below. The rain forests are disappearing in Australia, Hawaii, and South America. You see dust bowls developing in Africa and the Near East. Remote sensing devices, judging the humidity in the desert, have already warned you there will be plagues of locusts this year. To your amazement, you identify the lights of Denver and Cairo. And though you were taught about them one by one, as separate parts of a jigsaw puzzle, now you can see that the oceans, the atmosphere, and the land are not separate at all, but part of an intricate, recombining web of nature. Like Dorothy in *The Wizard of Oz*, you want to click your magic shoes together and say three times: "There's no place like home."

You know what home is. For many years, you've tried to be a modest and eager watcher of the skies, and of the Earth, whose green anthem you love. Home is a pigeon strutting like a petitioner in the courtyard in front of your house. Home is the law-abiding hickories out back. Home is the sign on a gas station just outside Pittsburgh that reads "If we can't fix it, it ain't broke." Home is springtime on campuses all across America, where students sprawl on the grass like the war-wounded at Gettysburg. Home is the Guatemalan jungle, at times deadly as an arsenal. Home is the pheasant barking hoarse threats at the neighbor's dog. Home is the exquisite torment of love and all the lesser mayhems of the heart. But what you long for is to stand back and see it whole. You want to live out that age-old yearning, portrayed in myths and legends of every culture, to step above the Earth and see the whole world fidgeting and blooming below you.

I remember my first flying lesson, in the doldrums of summer in upstate New York. Pushing the throttle forward, I zoomed down the runway until the undercarriage began to dance; then the ground fell away below and I was airborne, climbing up an invisible *flight* of stairs. To my amazement, the horizon came with me (how could it not on a round planet?). For the first time in my life I understood what a valley was, as I floated above one at 7,000 feet. I could see plainly the devastation of the gypsy moth, whose hunger had leeched the forests to a mottled gray. Later on, when I flew over Ohio, I was saddened to discover the stagnant ocher of the air, and to see that the long expanse of the Ohio River, dark and chunky, was the wrong texture for water, even flammable at times, thanks to the fumings of plastics factories, which I could also see, standing like pustules along the river. I began to understand how people settle a landscape, in waves and at crossroads, how they survey a land and irrigate it. Most of all, I discovered that there are things one can learn about the world only from certain perspectives. How can you understand the oceans without becoming part of its intricate fathoms? How can you understand the planet without walking upon it, sampling its marvels one by one, and then floating high above it, to see it all in a single eye-gulp?

Most of all, the twentieth century will be remembered as the time when we first began to understand what our address was. The "big, beautiful, blue, wet ball" of recent years is one way to say it. But a more profound way will speak of the orders of magnitude of that bigness, the shades of that blueness, the arbitrary delicacy of beauty itself, the ways in which water has made life possible, and the fragile euphoria of the complex ecosystem that is Earth, an Earth on which, from space, there are no visible fences, or military zones, or national borders. We need to send into space a flurry of artists and naturalists, photographers and painters, who will turn the mirror upon ourselves and show us Earth as a single planet, a single organism that's buoyant, fragile, blooming, buzzing, full of spectacles, full of fascinating human beings, something to cherish. Learning our full address may not end all wars, but it will enrich our sense of wonder and pride. It will remind us that the human context is not tight as a noose, but large as the universe we have the privilege to inhabit. It will change our sense of what a neighborhood is. It will persuade us that we are citizens of something larger and more profound than mere countries, that we are citizens of Earth, her joyriders and her caretakers, who would do well to work on her problems together. The view from space is offering us the first chance we evolutionary toddlers have had to cross the cosmic street and stand facing our own home, amazed to see it clearly for the first time.

Synesthesia

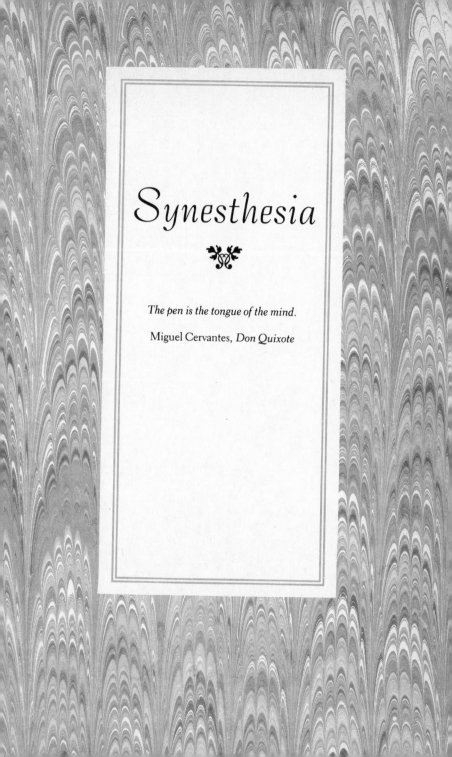

The pen is the tongue of the mind.

Miguel Cervantes, *Don Quixote*

FANTASIA

A creamy blur of succulent blue sound smells like week-old strawberries dropped onto a tin sieve as mother approaches in a halo of color, chatter, and a perfume like thick golden butterscotch. Newborns ride on intermingling waves of sight, sound, touch, taste, and, especially, smell. As Daphne and Charles Maurer remind us in *The World of the Newborn:*

> His world smells to him much as our world smells to us, but he does not perceive odors as coming through his nose alone. He hears odors, and sees odors, and feels them too. His world is a mêlée of pungent aromas—and pungent sounds, and bitter-smelling sounds, and sweet-smelling sights, and sour-smelling pressures against the skin. If we could visit the newborn's world, we would think ourselves inside a hallucinogenic perfumery.

In time, the newborn learns to sort and tame all its sensory impressions, some of which have names, many of which will remain nameless to the end of its days. Things that elude our verbal grasp are hard to pin down and almost impossible to remember. A cozy blur in the nursery vanishes into the rigorous categories of common sense. But for some people, that sensory blending never quits, and they taste baked beans whenever they hear the word "Francis," as one woman reported, or see yellow on touching a matte surface, or smell the passage of time. The stimulation of one sense stimulates another: *synesthesia* is the technical name, from the Greek *syn* (together) + *aisthanesthai* (to perceive). A thick garment of perception is woven thread by overlapping thread. A similar word is *synthesis*, in which the garment of thought is woven together idea by idea, and which originally referred to the light muslin clothing worn by the ancient Romans.

Daily life is a constant onslaught on one's perceptions, and everyone experiences some intermingling of the senses. According to Gestalt psychologists, when people are asked to relate a list of nonsense words to shapes and colors they identify certain sounds with certain shapes in ways that fall into clear patterns. What's more surprising is that this is true whether they are from the United States, England, the Mahali peninsula, or Lake Tanganyika. People with intense synesthesia tend to respond in predictable ways, too. A survey of two thousand synesthetes from various cultures revealed many similarities in the colors they assigned to sounds. People often associate low sounds with dark colors and high sounds with bright colors, for instance. A certain amount of synesthesia is built into our senses. If one wished to create instant synesthesia, a dose of mescaline or hashish would do nicely by exaggerating the neural connections between the senses. Those who experience intense synesthesia naturally on a regular basis are rare—only about one in every five hundred thousand people—and neurologist Richard Cytowič traces the phenomenon to the limbic system, the most primitive part of the brain, calling synesthetes "living cognitive fossils," because they may be people whose limbic system is not entirely governed by the much more sophisticated (and more recently evolved) cortex. As he says, "synesthesia . . . may be a memory of how early mammals saw, heard, smelled, tasted and touched."

While synesthesia drives some people to distraction, it drives distractions away from others. While it is a small plague to the person who doesn't want all that sensory overload, it invigorates those who are indelibly creative. Some of the most famous synesthetes have been artists. Composers Aleksandr Scriabin and Nikolai Rimski-Korsakov both freely associated colors with music when they wrote. To Rimski-Korsakov, C major was white; to Scriabin it was red. To Rimski-Korsakov, A major was rosy, to Scriabin it was green. More surprising is how closely their music-color synesthesias matched. Both associated E major with blue (for Rimski-Korsakov, it was sapphire blue, for Scriabin blue-white), A-flat major with purple (for Rimski-Korsakov it was grayish-violet, for Scriabin purple-violet), D major with yellow, etc.

Either writers have been especially graced with synesthesia, or they've been keener to describe it. Dr. Johnson once said that scarlet "represented nothing so much as the clangour of a trumpet." Baudelaire took pride in his sensory Esperanto, and his sonnet on the correspondences between perfumes, colors, and sounds greatly influenced the synesthesia-loving Symbolist movement. Symbol comes from the Greek word *symballein*, "to throw together," and, as *The Columbia Dictionary of Modern European Literature* explains, the Symbolists believed that "all arts are parallel translations of one fundamental mystery. Senses correspond to each other; a sound can be translated through a perfume and a perfume through a vision. . . . Haunted by these horizontal correspondances" and using suggestion rather than straightforward communication, they sought "the One hidden in Nature behind the Many." Rimbaud, who assigned colors to each of the vowel sounds and once described *A* as a "black hairy corset of loud flies," claimed that the only way an artist can arrive at life's truths is by experiencing "every form of love, of suffering, of madness," to be prepared for by "a long immense planned disordering of all the senses." The Symbolists, who were avid drug takers, delighted in the way hallucinogens intensified all their senses simultaneously. They would have loved (for a short time) taking LSD while watching Walt Disney's *Fantasia*, in which pure color dramatizes, melts into, and spurts from classical music. Few artists have written about synesthesia with the all-out precision and charm of Vladimir Nabokov, who, in *Speak, Memory*, analyzes what he calls his "colored hearing":

> Perhaps "hearing" is not quite accurate, since the color sensation seems to be produced by the very act of my orally forming a given letter while I imagine its outline. The long *a* of the English alphabet . . . has for me the tint of weathered wood, but a French *a* evokes polished ebony. This black group also includes hard *g* (vulcanized rubber) and *r* (a sooty rag being ripped). Oatmeal *n*, noodle-limp *l*, and the ivory-backed hand mirror of *o* take care of

the whites. I am puzzled by my French *on* which I see as the brimming tension-surface of alcohol in a small glass. Passing on to the blue group, there is steely *x,* thundercloud *z,* and huckleberry *k.* Since a subtle interaction exists between sound and shape, I see *q* as browner than *k,* while *s* is not the light blue of *c,* but a curious mixture of azure and mother-of-pearl. Adjacent tints do not merge, and diphthongs do not have special colors of their own, unless represented by a single character in some other language (thus the fluffy-gray, three-stemmed Russian letter that stands for *sh,* a letter as old as the rushes of the Nile, influences its English representation). . . . The word for rainbow, a primary, but decidedly muddy, rainbow, is in my private language the hardly pronounceable: *kzspygu.* The first author to discuss *audition colorée* was, as far as I know, an albino physician in 1812, in Erlangen.

The confessions of a synesthete must sound tedious and pretentious to those who are protected from such leaking and drafts by more solid walls than mine are. To my mother, though, this all seemed quite normal. The matter came up, one day in my seventh year, as I was using a heap of old alphabet blocks to build a tower. I casually remarked to her that their colors were all wrong. We discovered then that some of her letters had the same tint as mine and that, besides, she was optically affected by musical notes. These evoked no chromatisms in me whatsoever.

Synesthesia can be hereditary, so it's not surprising that Nabokov's mother experienced it, nor that it expressed itself slightly differently in her son. However, it's odd to think of Nabokov, Faulkner, Virginia Woolf, Huysmans, Baudelaire, Joyce, Dylan Thomas and other notorious synesthetes as being more primitive than most people, but that may indeed be true. Great artists feel at home in the luminous spill of sensation, to which they add their own complex sensory Niagara. It would certainly have amused Nabokov to imagine himself closer than others to his mammalian ancestors, which he would no doubt have depicted in a fictional hall of mirrors with suave, prankish, Nabokovian finesse.

COURTING THE MUSE

What a strange lot writers are, we questers after the perfect word, the glorious phrase that will somehow make the exquisite avalanche of consciousness sayable. We who live in mental barrios, where any roustabout idea may turn to honest labor, if only it gets the right incentive—a bit of drink, a light flogging, a delicate seduction. I was going to say that our heads are our offices or charnel houses, as if creativity lived in a small walk-up flat in Soho. We know the mind doesn't dwell in the brain alone, so the where of it is as much a mystery as the how. Katherine Mansfield once said that it took "terrific hard gardening" to produce inspiration, but I think she meant something more willful than Picasso's walks in the forests of Fontainebleau, where he got an overwhelming "indigestion of greenness," which he felt driven to empty onto a canvas. Or maybe that's exactly what she meant, the hard gardening of knowing where and when and for how long and precisely in what way to walk, and then the will to go out and walk it as often as possible, even when one is tired or isn't in the mood, or has only just walked it to no avail. Artists are notorious for stampeding their senses into duty, and they've sometimes used remarkable tricks of synesthesia.

Dame Edith Sitwell used to lie in an open coffin for a while before she began her day's writing. When I mentioned this macabre bit of gossip to a poet friend, he said acidly: "If only someone had thought to shut it." Picture Dame Edith, rehearsing the posture of the grave as a prelude to the sideshows on paper she liked to stage. The straight and narrow was never her style. Only her much-ridiculed nose was rigid, though she managed to keep it entertainingly out of joint for most of her life. What was it exactly about that dim, contained solitude that spurred her creativity? Was it the idea of the coffin or the feel, smell, foul air of it that made creativity possible?

Edith's horizontal closet trick may sound like a prank unless you look at how other writers have gone about courting their muses. The poet Schiller used to keep rotten apples under the lid of his desk and inhale their pungent bouquet when he needed to find the right word. Then he would close the drawer, although the fragrance remained in his head. Researchers at Yale University discovered that the smell of spiced apples has a powerful elevating effect on people and can even stave off panic attacks. Schiller may have sensed this all along. Something in the sweet, rancid mustiness of those apples jolted his brain into activity while steadying his nerves. Amy Lowell, like George Sand, enjoyed smoking cigars while writing, and in 1915 went so far as to buy 10,000 of her favorite Manila stogies to make sure she could keep her creative fires kindled. It was Lowell who said she used to "drop" ideas into her subconscious "much as one drops a letter into the mailbox. Six months later, the words of the poem began to come into my head. . . . The words seem to be pronounced in my head, but with nobody speaking them." Then they took shape in a cloud of smoke. Both Dr. Samuel Johnson and the poet W. H. Auden drank colossal amounts of tea—Johnson was reported to have frequently drunk twenty-five cups at one sitting. Johnson did die of a stroke, but it's not clear if this was related to his marathon tea drinking. Victor Hugo, Benjamin Franklin, and many others felt that they did their best work if they wrote in the nude. D. H. Lawrence once even confessed that he liked to climb naked up mulberry trees—a fetish of long limbs and rough bark that stimulated his thoughts.

Colette used to begin her day's writing by first picking fleas from her cat, and it's not hard to imagine how the methodical stroking and probing into fur might have focused such a voluptuary's mind. After all, this was a woman who could never travel light, but insisted on taking a hamper of such essentials as chocolate, cheese, meats, flowers, and a baguette whenever she made even brief sorties. Hart Crane craved boisterous parties, in the middle of which he would disappear, rush to a typewriter, put on a record of a Cuban rumba, then Ravel's *Boléro,* then a torch song, after which he would return, "his face brick-red, his eyes burning, his already iron-gray hair straight up from his skull. He would be chewing a five-cent cigar which he had forgotten to light. In his hands would be two or three sheets of typewritten manuscript. . . . 'Read that,' he would say, 'isn't that the *grrrea*test poem ever written!' " This is Malcolm Cowley's account, and Cowley goes on to offer even more examples of how Crane reminded him of "another friend, a famous killer of woodchucks," when the writer "tried to charm his inspiration out of its hiding place by drinking and laughing and playing the phonograph."

Stendhal read two or three pages of the French civil code every morning before working on *The Charterhouse of Parma*—"in order" he said, "to acquire the correct tone." Willa Cather read the Bible. Alexandre Dumas *père* wrote his nonfiction on rose-colored paper, his fiction on blue, and his poetry on yellow. He was nothing if not orderly, and to cure his insomnia and regularize his habits he went so far as to eat an apple at seven each morning under the Arc de Triomphe. Kipling demanded the blackest ink he could find and fantasized about keeping "an ink-boy to grind me Indian ink," as if the sheer weight of the blackness would make his words as indelible as his memories.

Alfred de Musset, George Sand's lover, confided that it piqued him when she went directly from lovemaking to her writing desk, as she often did. But surely that was not so direct as Voltaire, who used his lover's naked back as a writing desk. Robert Louis Stevenson, Mark Twain, and Truman Capote all used to lie down when they wrote, with Capote going so far as to declare himself "a completely horizontal writer." Writing students often hear that Hemingway wrote standing up, but not that he obsessively sharpened pencils first, and, in any case, he wasn't standing up out of some sense of himself as the sentinel of tough, ramrod prose, but because he had hurt his back in a plane crash. Poe supposedly wrote with his cat sitting on his shoulder. Thomas Wolfe, Virginia Woolf, and Lewis Carroll were all standers; and Robert Hendrickson reports in *The Literary Life and Other Curiosities* that Aldous Huxley "often wrote with his nose." In *The Art of Seeing,* Huxley says that "a little nose writing will result in a perceptible temporary improvement of defective vision."

Many nonpedestrian writers have gotten their inspiration from walking. Especially poets—there's a sonneteer in our chests; we walk around to the beat of iambs. Wordsworth, of course, and John Clare, who used to go out looking for the horizon and one day in insanity thought he found it, and A. E. Housman, who, when asked to define poetry, had the good sense to say: "I could no more define poetry than a terrier can a rat, but I thought we both recognized the object by the symptoms which it provokes in us. . . . If I were obliged . . . to name the class of things to which it belongs, I should call it a secretion." After drinking a pint of beer at lunch, he would go out for a two- or three-mile walk and then gently secrete.

I guess the goal of all these measures is concentration, that petrified mirage, and few people have written about it as well as Stephen Spender did in his essay "The Making of a Poem":

> There is always a slight tendency of the body to sabotage the attention of the mind by providing some distraction. If this need for distraction can be directed into one channel—such as the odor of rotten apples or the taste of tobacco or tea—then other distractions outside oneself are put out of the competition. Another possible explanation is that the concentrated effort of writing poetry is a spiritual activity which makes one completely forget, for the time being, that one has a body. It is a disturbance of the balance of the body and mind and for this reason one needs a kind of anchor of sensation with the physical world.

This explains, in part, why Benjamin Franklin, Edmond Rostand, and others wrote while soaking in a bathtub. In fact, Franklin brought the first bathtub to the United States in the 1780s and he loved a good, long, thoughtful submersion. In water and ideas, I mean. Ancient Romans found it therapeutic to bathe in asses' milk or even in crushed strawberries. I have a pine plank that I lay across the sides of the tub so that I can stay in a bubble bath for hours and write. In the bath, water displaces much of your weight, and you feel light, your blood pressure drops. When the water temperature and the body temperature converge, my mind lifts free and travels by itself. One summer, lolling in baths, I wrote an entire verse play, which mainly consisted of dramatic monologues spoken by the seventeenth-century Mexican poet Sor Juana Ínez de la Cruz; her lover, an Italian courtier; and various players in her tumultuous life. I wanted to slide off the centuries as if from a hill of shale. Baths were perfect.

The Romantics, of course, were fond of opium, and Coleridge freely admitted to indulging in two grains of it before working. The list of writers triggered to inspirational highs by alcohol would occupy a small, damp book. T. S. Eliot's tonic was viral—he preferred writing when he had a head cold. The rustling of his head, as if full of petticoats, shattered the usual logical links between things and allowed his mind to roam.

Many writers I know become fixated on a single piece of music when they are writing a book, and play the same piece of music perhaps a thousand times in the course of a year. While he was writing the novel *The Place in Flowers Where Pollen Rests,* Paul West listened nonstop to sonatinas by Ferruccio Busoni. He had no idea why. John Ashbery first takes a walk, then brews himself a cup of French blend Indar tea, and listens to something post-Romantic ("the chamber music of Franz Schmidt has been beneficial" he told me). Some writers become obsessed with cheap and tawdy country-and-western songs, others with one special prelude or tone poem. I think the music they choose creates a mental frame around the essence of the book. Every time the music plays, it re-creates the emotional terrain the writer knows the book to live in. Acting as a mnemonic of sorts, it guides a fetishistic listener to the identical state of alert calm, which a brain-wave scan would probably show.

When I asked a few friends about their writing habits, I thought for sure they'd fictionalize something offbeat—standing in a ditch and whistling Blake's "Jerusalem," perhaps, or playing the call to colors at Santa Anita while stroking the freckled bell of a foxglove. But most swore they had none—no habits, no superstitions, no special routines. I phoned William Gass and pressed him a little.

"You have no unusual work habits?" I asked, in as level a tone as I could muster. We had been colleagues for three years at Washington University, and I knew his quiet professorial patina concealed a truly exotic mental grain.

"No, sorry to be so boring," he sighed. I could hear him settling comfortably on the steps in the pantry. And, as his mind is like an overflowing pantry, that seemed only right.

"How does your day begin?"

"Oh, I go out and photograph for a couple of hours," he said.

"What do you photograph?"

"The rusty, derelict, overlooked, downtrodden parts of the city.

Filth and decay mainly," he said in a nothing-much-to-it tone of voice, as casually dismissive as the wave of a hand.

"You do this every day, photograph filth and decay?"

"Most days."

"And then you write?"

"Yes."

"And you don't think this is unusual?"

"Not for me."

A quiet, distinguished scientist friend, who has published two charming books of essays about the world and how it works, told me that his secret inspiration was "violent sex." I didn't inquire further, but noted that he looked thin. The poets May Swenson and Howard Nemerov both told me that they like to sit for a short spell each day and copy down whatever pours through their heads from "the Great Dictator," as Nemerov labels it, then plow through to see what gems may lie hidden in the rock. Amy Clampitt, another poet, told me she searches for a window to perch behind, whether it be in the city or on a train or by the seaside. Something about the petri dish effect of the glass clarifies her thoughts. The novelist Mary Lee Settle tumbles out of bed and heads straight for her typewriter, before the dream state disappears. Alphonso Lingis—whose unusual books, *Excesses* and *Libido,* consider the realms of human sensuality and kinkiness—travels the world sampling its exotic erotica. Often he primes the pump by writing letters to friends. I possess some extraordinary letters, half poetry, half anthropology, he sent me from a Thai jail (where he took time out from picking vermin to write), a convent in Ecuador, Africa (where he was scuba-diving along the coast with filmmaker Leni Riefenstahl), and Bali (where he was taking part in fertility rituals).

Such feats of self-rousing are awkward to explain to one's parents, who would like to believe that their child does something reasonably normal, and associates with reasonably normal folk, not people who sniff rotten apples and write in the nude. Best not to tell them how the painter J. M. W. Turner liked to be lashed to the mast of a ship and taken sailing during a real hell-for-leather storm so that he could be right in the middle of the tumult. There are many roads to Rome, as the old maxim has it, and some of them are sinewy and full of fungus and rocks, while others are paved and dull. I think I'll tell my parents that I stare at bouquets of roses before I work. Or, better, that I stare at them until butterflies appear. The truth is that, besides opening and closing mental drawers (which I picture in my mind), writing in the bath, beginning each summer day by choosing and arranging flowers for a Zenlike hour or so, listening obsessively to music (Alessandro Marcello's oboe concerto in D minor, its adagio, is what's nourishing my senses at the moment), I go speed walking for an hour every single day. Half of the oxygen in the state of New York has passed through my lungs at one time or another. I don't know whether this helps or not. My muse is male, has the radiant silvery complexion of the moon, and never speaks to me directly.

POSTSCRIPT

There is a point beyond which the senses cannot lead us. Ecstasy means being flung out of your usual self, but that is still to feel a commotion inside. Mysticism transcends the here and now for loftier truths unexplainable in the straitjacket of language; but such transcendence registers on the senses, too, as a rush of fire in the veins, a quivering in the chest, a quiet, fossillike surrender in the bones. Out-of-body experiences aim to shed the senses, but they cannot. One may see from a new perspective, but it's still an experience of vision. Computers now help to interpret some of life's processes, which we previously used only our senses to seek, trace, and understand. Astronomers are more apt to look at their telescope's monitors than to consider the stars with their naked eyes. But we continue to use our senses to interpret the work of the computers, to see the monitors, to judge and analyze, and to design ever newer dreams of artificial intelligence. Never will we leave the palace of our perceptions.

If we are in a rut, it is a palatial and exquisite rut. And yet, like prisoners in a cell, we grip our ribs from within, rattle them, and beg for release. In the Bible, God instructs Moses to burn incense sweet and to His liking. Does God have nostrils? How can a god prefer one smell of this earth to another? The rudiments of decay complete a cycle necessary for growth and deliverance. Carrion smells offensive to us, but delicious to those animals who rely on it for food. What they excrete will make the soil rich and the crops abundant. There is no need for divine election. Perception is itself a form of grace.

In 1829, Goethe, writing about color theory, said: "One searches in vain beyond phenomenon; it in itself is revelation."

There is so much physical variation among people—some have strong hearts, some have weak bladders, some have steadier hands than others, some have bad eyesight—it's only logical that senses should vary, too. Yet how much in agreement our senses are—so much so that scientists can define a "red wave" by saying that it is produced by a vibration of 660 millimicrons, which stimulates the retinas to see red. Tones are defined equally precisely, as are the temperatures at which we feel hot or cold. Our senses unite us in a common field of temporal glory, but they can also divide us. Sometimes briefly, or, as in the case of artists, for a lifetime.

I woke one morning this winter after a sudden heavy snowfall to see the evergreens in front of my house bent in half under a burden of snow and ice. Unless I freed them, they would snap under their own weight, so I took a shovel and started bashing the branches to shake the snow down. Suddenly one of the heaviest branches let fly, and snow burned my face like sunlight, iced and clung and kept on pouring as I stood, chin tilted toward the dam-burst, pillar-calm, with my every sense alert. But what a puzzle for the neighbor boy, jarred from his play by that basso *whump!*, to see a madwoman gripped by her own storm. Out of the corner of my eye, I saw him wrinkle his face, then ravel his sled-tow and tramp away. For me, time did a lazy soft-shoe; long minutes seemed to pass, and I thought of mammoths, goose down, Ice-Age cunning, the long white drawl of a glacier on the move, snow avalanching down a polar chasm. For him, the same moment fled like a gnat.

For convenience, and perhaps in a kind of mental pout about how thickly demanding just being alive is, we say there are five senses. Yet we know there are more, should we but wish to explore and canonize them. People who dowse for water are probably responding to an electromagnetic sense we all share to a greater or lesser degree. Other animals, such as butterflies and whales, navigate in part by reading the earth's magnetic fields. It wouldn't surprise me to learn that we, too, have some of that magnetic awareness. We were nomads for so much of our history. We are as phototropic as plants, smitten with the sun's light, and this should be considered a sense separate from vision, with which it has little to do. Our experience of pain is quite different from the other worlds of touch. Many animals have infrared, heat-sensing, electromagnetic, and other so-phisticated ways of perceiving. The praying mantis uses ultrasonics to communicate. Both the alligator and the elephant use infrasonics. The duckbill platypus swings its bill back and forth underwater, using it as an antenna to pick up electrical signals from the muscles of the crustaceans, frogs, and small fish on which it preys. The vibratory sense, so highly developed in spiders, fish, bees, and other animals, needs to be studied more in human beings. We have a muscular sense that guides us when we pick up objects—we know at once that they are heavy, light, solid, hard, or soft, and we can figure out how much pressure or resistance will be required. We are constantly aware of a sense of gravity, which counsels us about which way is up and how to rearrange our bodies if we're falling, or climb-ing, or swimming, or bent at some unusual angle. There is the proprioceptive sense, which tells us what position each component of our bodies is in at any moment in our day. If the brain didn't always know where the knees or the lungs were, it would be impossi-ble to walk or breathe. There seems to be a complex space sense that, as we move into an era of space stations and cities and lengthy space travel, we will need to understand in detail. Prolonged Earthlessness alters our physiology and also the evidence of our senses, in part because of the rigors of being in zero gravity,* and in part because of the lidless sprawl of deep space itself, in which there are few sensory handrails, guides or landmarks, and everywhere you look there is not scene but pure vista.

Species evolve senses fine-tuned for different programs of survival, and it's impossible to put ourselves into the sensory realm of any other species. We've evolved unique human ways of perceiving the world to cope with the demands of our environment. Physics sets the limits, but biology and natural selection determine where an animal

---

*For example, the face swells as body fluids drift upward, and the brain signals the body to get rid of this excess fluid by urinating more and drinking less.

will fall among all the sensory possibilities. When scientists, philosophers, and other commentators speak of the real world, they're talking about a myth, a convenient fiction. The world is a construct the brain builds based on the sensory information it's given, and the information is only a small part of all that's available. We can modify our senses through bat detectors, binoculars, telescopes, and microscopes, broadening that sensory horizon, and there are instruments that allow us to become a kind of sensory predator that natural selection never meant us to be. Physicists explain that molecules are always moving: The book in front of you is actually squirming under your fingertips. But we don't see this motion at that molecular level, because it's not evolutionarily important that we do. We're given only the sensory information crucial to our survival.

Evolution didn't overload us with unnecessary abilities. For example: We may use numbers in the millions and trillions, but they are basically meaningless to us. Many things are unavailable to us because they're not part of our distant evolutionary background. In an odd way, one-celled animals may have a more realistic sense of the world than higher animals do, because they respond to every stimulus they encounter. We, on the other hand, select only a few. The body edits and prunes experience before sending it to the brain for contemplation or action. Not every whim of the wind triggers the hair on the wrist to quiver. Not every vagary of sunlight registers on the retina. Not everything we feel is felt powerfully enough to send a message to the brain; the rest of the sensations just wash over us, telling us nothing. Much is lost in translation, or is censored, and in any case our nerves don't all fire at once. Some of them remain silent, while others respond. This makes our version of the world somewhat simplistic, given how complex the world is. The body's quest isn't for truth, it's for survival.

Our senses also crave novelty. Any change alerts them, and they send a signal to the brain. If there's no change, no novelty, they doze and register little or nothing. The sweetest pleasure loses its thrill if it continues too long. A constant state—even of excitement—in time becomes tedious, fades into the background, because our senses have evolved to report changes, what's new, something startling that has to be appraised: a morsel to eat, a sudden danger. The body takes stock of the world like an acute and observant general moving through a complex battleground, looking for patterns and stratagems. So it is not only possible but inevitable that a person will grow used to a city's noises and visual commotion and not register these stimuli constantly. On the other hand, novelty itself will always rivet one's attention. There is that unique moment when one confronts something new and astonishment begins. Whatever it is, it looms brightly, its edges sharp, its details ravishing, in a hard clear light; just beholding it is a form of revelation, a new sensory litany. But the second time one sees it, the mind says, Oh, that again, another wing walker, another moon landing. And soon, when it's become commonplace, the brain begins slurring the details, recognizing it too quickly, by just a few of its features; it doesn't have to bother scrutinizing it. Then it is lost to astonishment, no longer an extraordinary instance but a generalized piece of the landscape. Mastery is what we strive for, but once we have it we lose the precarious superawareness of the amateur. "It's old hat," we say, as if such an old, weatherbeaten article of clothing couldn't yield valuable insights about its wearer and the era in which it was created and crushed. "Old news," we say, even if the phrase is an oxymoron. News is new and should sound an alarm in our minds. When it becomes old, what happens to its truth? "He's history," we say, meaning that someone is no longer new for us, no longer fresh and stimulating, but banished to the world of fossil and ruin. So much of our life passes in a comfortable blur. Living on the senses requires an easily triggered sense of marvel, a little extra energy, and most people are lazy about life. Life is something that happens to them while they wait for death. Many millennia from now, will we evolve into people who will perceive the world differently, employ the senses differently, and perhaps know the world more intimately? Or will those future souls, perhaps further away from any physical sense of the world, envy us, the passionate and thrill-seeking ones, who gorged ourselves on life, sense by sense, dream by dream?

Hold a glance a little longer than usual, let the eyes smolder and a smile creep onto the lips, and a small toboggan run forms in the chest as the heart gets ready to race. Novelty plays a large role in sexual arousal, as e. e. cummings, a master of sensuality and titivation, suggests in his poem "96":

> i like my body when it is with your
> body. It is so quite new a thing.
> Muscles better and nerves more.
> i like your body. i like what it does,
> i like its hows. i like to feel the spine
> of your body and its bones, and the trembling
> -firm-smooth ness and which i will
> again and again and again
> kiss, i like kissing this and that of you,
> i like, slowly stroking the shocking fuzz
> of your electric fur, and what-is-it comes
> over parting flesh. . . . And eyes big love-crumbs,
>
> and possibly i like the thrill
>
> of under me you so quite new

When cummings wrote this beautiful love sonnet, he certainly didn't know (or need to) that studies would later reveal how men's testosterone levels jump when a new woman enters the room. The simple fact of her novelty is physically exciting. But the same is true for women and their hormones when a new man enters the room. For social, moral, esthetic, parental, religious, or even mystical reasons, we may choose to live with one partner for life, but our instincts nag at us. There is nothing like the thrill of being new for someone. And even though everything related to love—the roller coaster of flirtation, the thrust and parry of courtship, the razzle-dazzle of lovemaking—has probably evolved so that two people who have a good chance of producing and raising hearty offspring will find each other and mate with a strong biological sense of purpose, we don't always feel obliged to play by nature's rules. The challenge (and highwire fun) of love is finding ways to make each day a fresh adventure with one's partner.

Life teaches us to be guarded. We use words like *vulnerable* when we mean that we are letting down a drawbridge over the moat of our self-protection and trusting another inside the fortress of our lives. Lovers combine their senses, blend their electrical impulses, help sense for one another. When they touch, their bodies double in size. They get under each other's skin, literally and emotionally. During intercourse, a man hides part of himself in a woman, a bit of his body disappears from view, while a woman opens up the internal workings of her body and adds another organ to it, as if it were meant to be there all along. These, in a starched, stiff, dangerous world, are ultimate risks.

But suppose you could sense any world you wanted to? At NASA's Ames Research Center, in Mountain View, California, researchers have been perfecting "Virtual Reality" garb—a mask and gloves that extend one's senses, which are, both in appearance and power, reminiscent of the magic regalia heroes sometimes relied on in epic sagas. Don the sensor-equipped gloves, and you can reach into a computer-generated landscape and move things around. Wear the mask, and you can see an invisible or imaginary world as if it were perfectly viewable, full of depth and color—it might be the rolling sand dunes of Mars, or an approach to O'Hare Airport in fog, or perhaps a faulty space station generator. Why watch a murder mystery from across a room, when you can put on a mask and glove and walk right into the action and handle the clues. How could such a sleight of hand, mind, mask, and senses be possible?

One of the most profound paradoxes of being human is that the thick spread of sensation we relish isn't perceived directly by the brain. The brain is silent, the brain is dark, the brain tastes nothing, the brain hears nothing. All it receives are electrical impulses—not the sumptuous chocolate melting sweetly, not the oboe solo like the flight of a bird, not the tingling caress, not the pastels of peach and lavender at sunset over a coral reef—just impulses. The brain is blind, deaf, dumb, unfeeling. The body is a transducer (from Latin, *transducere,* to lead across, transfer), a device that converts energy of one sort to energy of another sort, and that is its genius. Our bodies take mechanical energy and convert it to electrical energy. I touch the soft petal of a red rose called "Mr. Lincoln," and my receptors translate that mechanical touch into electrical impulses that the brain reads as soft, supple, thin, curled, dewy, velvety: rose petal–like. When Walt Whitman said: "I sing the body electric," he didn't know how prescient he was. The body does indeed sing with electricity, which the mind deftly analyzes and considers. So, to some extent, reality is an agreed-upon fiction. How silly, then, that philosophers should quarrel about appearance and reality. The universe will be knowable to other creatures in other ways.

A dolphin has a brain as complex as our own; it has language, culture, and emotions. It has its own society, with codes of conduct, family groups, and a civilization, but it lives in a world on "our" planet, as we like to say with chauvinistic bravado, unimaginably different from our own. We may have much to learn from it. Deep down, we know our devotion to reality is just a marriage of convenience, and we leave it to the seers, the shamans, the ascetics, the religious teachers, the artists among us to reach a higher state of awareness, from which they transcend our rigorous but routinely analyzing senses and become closer to the raw experience of nature that pours into the unconscious, the world of dreams, the source of myth. "How do you know but that every bird that cleaves the aerial way is not an immense world of delight closed to your senses five?" William Blake wrote. We have much to learn from and about the senses of animals. Otherwise, how shall we hope to be good caretakers of the planet, should that turn out to be our role? How shall we appreciate our small part in the web of life on Earth? How shall we understand the minds of extraterrestrials, if we make contact with them? How shall we come to know one another deeply, compassionately, fulfillingly, unless we learn more of how the mind and senses work? Our several senses, which feel so personal and impromptu, and seem at times to divorce us from other people, reach far beyond us. They're an extension of the genetic chain that connects us to everyone who has ever lived; they bind us to other people and to animals, across time and country and happenstance. They bridge the personal and the impersonal, the one private soul with its many relatives, the individual with the universe, all of life on Earth. In REM sleep, our brain waves range between eight and thirteen hertz, a frequency at which flickering light can trigger epileptic seizures. The tremulous earth quivers gently at around ten hertz. So, in our deepest sleep, we enter synchrony with the trembling of the earth. Dreaming, we become the Earth's dream.

It began in mystery, and it will end in mystery. However many of life's large, captivating principles and small, captivating details we may explore, unpuzzle, and learn by heart, there will still be vast unknown realms to lure us. If uncertainty is the essence of romance, there will always be enough uncertainty to make life sizzle and renew our sense of wonder. It bothers some people that no matter how passionately they may delve, the universe remains inscrutable. "For my part," Robert Louis Stevenson once wrote, "I travel not to go anywhere, but to go. I travel for travel's sake. The great affair is to move." The great affair, the love affair with life, is to live as variously as possible, to groom one's curiosity like a high-spirited thoroughbred, climb aboard, and gallop over the thick, sun-struck hills every day. Where there is no risk, the emotional terrain is flat and unyielding, and, despite all its dimensions, valleys, pinnacles, and detours, life will seem to have none of its magnificent geography, only a length. It began in mystery, and it will end in mystery, but what a savage and beautiful country lies in between.

FURTHER READING

GENERAL

Bachelard, Gaston. *The Poetics of Space*. Boston: Beacon Press, 1969.

Bates, H. E. *The Purple Plain*. London: Penguin Books, 1974.

Bodanis, David. *The Secret House*. New York: Simon & Schuster, Inc., 1986.

Bonner, John Tyler. *The Scale of Nature*. New York: Harper & Row, 1969.

Brash, R. *How Did It Begin? Supersitions and Their Romantic Origins*. Australia: Longmans, Green & Co., Ltd., 1965.

Braudel, Fernand. *The Structures of Everyday Life*. New York: Harper & Row, 1982.

Buddenbrock, Wolfgang von. *The Senses*. Ann Arbor, Michigan: The University of Michigan Press.

Campbell, Joseph. *The Power of Myth*. Betty Sue Flowers, ed., introduction by Bill Moyers. Garden City, New York: Doubleday, 1988.

Carcopino, Jerome. *Daily Life in Ancient Rome*. Harry T. Lowell, ed. New Haven, Connecticut: Yale University Press, 1940.

Carr, Donald E. *The Forgotten Senses*. Garden City, New York: Doubleday, 1972.

Dubkin, Leonard. *The White Lady*. London: Macmillan & Co., Ltd., 1952.

Eiseley, Loren. *The Immense Journey*. New York: Random House, Inc./Vintage Books, 1957.

———. *The Lost Notebooks of Loren Eiseley*. Kenneth Hever, ed. Boston: Little, Brown & Co., 1987.

Froman, Robert. *The Many Human Senses*. London: G. Bell and Sons, Ltd., 1966.

Gass, William. *On Being Blue*. Boston: Godine, 1976.

Glassner, Barry. *Bodies: Why We Look the Way We Do*. New York: G. P. Putnam's Sons, 1988.

Guiness, Alma E., ed. *ABC's of the Human Body*. Pleasantville, New York: Reader's Digest Books, 1987.

Huizinga, Johan. *Homo Ludens: A Study of the Play Element in Culture*. Boston: Beacon Press, 1955.

Huysmans, J.-K. *Against Nature*. New York: Penguin Books, 1986.

Lingis, Alphonso. *Excesses: Eros and Culture*. Albany, New York: State University of New York, 1978.

Maeterlinck, Maurice. *The Life of the Bee.* New York: New American Library, 1954.

Martin, Russell. *Matters Gray & White.* New York: Fawcett/Crest, 1986.

Milne, Lorus and Margery. *The Senses of Animals and Men.* New York: Atheneum, 1964.

Morris, Desmond. *Bodywatching.* New York: Crown, 1985.

———. *Catwatching.* New York: Crown, 1986.

———. *Dogwatching.* New York: Crown, 1987.

———. *Intimate Behavior.* New York: Bantam, 1973.

———. *Manwatching.* New York: Abrams, 1977.

Murchie, Guy. *The Seven Mysteries of Life: An Exploration in Science and Philosophy.* Boston: Houghton Mifflin Company, 1978.

Panati, Charles. *The Browser's Book of Beginnings.* Boston: Houghton Mifflin Company, 1984.

———. *Extraordinary Origins of Everyday Things.* New York: Harper & Row, 1987.

Parker, Arthur C. *Indian How Book.* New York: Dover, 1954.

Polhemus, Ted, ed. *The Body Reader: Social Aspects of the Human Body.* New York: Pantheon Books, 1978.

Poole, Robert M., ed. *The Incredible Machine.* Washington, D.C.: National Geographic Society, 1986.

Rilke, Rainer Maria, trans. G. Craig Houston. *Where Silence Reigns: Selected Prose.* New York: New Directions, 1978.

Rivlin, Robert, and Karen Gravelle. *Deciphering the Senses: The Expanding World of Human Perception.* New York: Simon & Schuster, 1984.

Robinson, Howard F., et al. *Colors in the Wild.* Washington, D.C.: National Wildlife Federation, 1985.

Sagan, Carl. *The Dragons of Eden.* New York: Random House, Inc., 1977.

Selzer, Richard. *Mortal Lessons.* New York: Simon & Schuster, 1976.

Smith, Anthony. *The Body.* New York: Penguin Books, 1986.

Thompson, D'Arcy W. *On Growth and Form.* Cambridge, Massachusetts: Cambridge University Press, 1961.

van der Post, Laurens. *The Heart of the Hunter.* New York: Harcourt Brace Jovanovich, 1980.

Von Frisch, Karl. *Animal Architecture.* New York: Harcourt Brace Jovanovich, 1974.

Walker, Stephen. *Animal Thoughts.* London: Routledge & Kegan Paul, Ltd., 1983.

Walsh, William S. *Curiosities of Popular Customs.* London: J. P. Lippincott Co., 1897.

Wilentz, Joan Steen. *The Senses of Man.* New York: Crowell, 1968.

Wilson, Edward O. *Biophilia.* Cambridge, Massachusetts: Harvard University Press, 1984.

SMELL

Bedichek, Roy. *The Sense of Smell*. Garden City, New York: Doubleday, 1960.

Bloch, Iwan. *Odoratus Sexualis*. New York: New York Anthropological Society, 1937.

Burton, Robert. *The Language of Smell*. London: Routledge & Kegan Paul, 1976.

Corbin, Alain. *The Foul and the Fragrant*. Cambridge, Massachusetts: Harvard University Press, 1986.

Erb, Russell C. *The Common Scents of Smell*. New York: World Publishing Co., 1968.

Ferenczi, Sandor. *Thalassa: A Theory of Genitality*. New York: W. W. Norton, 1968.

Gombrowicz, Witold. *Diary, Vol. I*. Evanston, Illinois: Northwestern University Press, 1988.

Harkness, Jack. *The Makers of Heavenly Roses*. London: Souvenir Press, 1985.

Moncrieff, R. W. *Odours*. London: William Heinemann Medical Books Ltd., 1970.

Morris, Edwin T. *Fragrance*. New York: Scribner's, 1986.

Muller, Julia, et al. *Fragrance Guide (Feminine Notes)*. London: Johnson Publications, n.d.

————, with Dr. Hans Brauer and Joachim Mensing. *The H & R Book of Perfume*. London: Johnson Publications, n.d.

Ray, Richard, and Michael MacCarkey. *Roses*. Tucson, Arizona: H. P. Books, 1981.

Süskind, Patrick. *Perfume*. New York: Alfred A. Knopf, Inc., 1987.

West, Paul. *The Place in Flowers Where Pollen Rests*. Garden City, New York: Doubleday, 1988.

TOUCH

Allen, J. W. T., ed. and trans. *The Customs of the Swahili People*. Berkeley and Los Angeles: University of California Press, 1981

BBC/WGBH. "A Touch of Sensitivity." December 9, 1980.

Beardsley, Timothy. "Benevolent Bradykinins." *Scientific American*, July 1988.

Fellman, Sandi, ed. *The Japanese Tattoo*. New York: Abbeville Press, 1987.

Gallico, G. Gregory, et al. "Permanent Coverage of Large Burn Wounds with Autologous Cultured Human Epithelium." *The New England Journal of Medicine*, Vol. 311, No. 7, August 16, 1984.

Goleman, Daniel. "The Experience of Touch: Research Points to a Critical Role." *The New York Times*, February 2, 1988, p. C1.

Lamb, Michael. "Second Thoughts on First Touch." *Psychology Today*, Vol. 16, No. 4, April 1982.

Lebeck, Robert. *The Kiss.* New York: St. Martin's Press, 1981.

Macrae, Janet. *Therapeutic Touch: A Practical Guide.* New York: Alfred A. Knopf, Inc., 1988.

Montagu, Ashley. *Touching: The Human Significance of the Skin.* New York: Columbia University Press, 1971.

Nyrop, Christopher, trans. W. F. Harvey. *The Kiss and Its History.* London: Sand and Co., 1901.

Perella, Nicolas James. *The Kiss Sacred and Profane.* Berkeley and Los Angeles: University of California Press, 1969.

Sachs, Frederick. "The Intimate Sense of Touch." *The Sciences,* January/February 1988.

TASTE

Angier, Bradford. *How to Stay Alive in the Woods.* New York: Macmillan, 1962.

Brillat-Savarin, Anthelme, trans. and annotated by M. F. K. Fisher. *The Physiology of Taste.* San Francisco, California: North Point Press, 1986.

Farb, Peter, and George Armelagos. *Consuming Passions.* New York: Washington Square Press, 1970.

Ferrary, Jeannette. "Plain Old Vanilla Isn't All that Plain Anymore." *The New York Times,* January 13, 1988.

Harris, Marvin. *The Sacred Cow and the Abominable Pig: Riddles of Food and Culture.* New York: Simon & Schuster/Touchstone Books, 1987.

Liebowitz, Michael. *The Chemistry of Love.* New York: Berkeley Books, 1984.

Pullar, Philippa. *Consuming Passions.* Boston: Little, Brown & Company, 1970.

Tisdale, Sallie. *Lot's Wife: Salt and the Human Condition.* New York: Henry Holt & Co., 1988.

HEARING

Attali, Jacques, trans. Brian Massumi. *Noise: The Political Economy of Music.* Minneapolis: University of Minnesota Press, 1985.

Bach, Johann Sebastian. *Complete Organ Works.* With a preface by Dr. Albert Schweitzer and Charles-Marie Widor. New York: G. Schirmer, Inc., 1912.

Broad, William J. "Complex Whistles Found to Play Key Roles in Inca and Maya Life." *The New York Times,* March 29, 1988.

Chatwin, Bruce. *The Songlines.* New York: The Viking Press, 1987.

Conniff, Richard. "When the Music in Our Parlors Brought Death to Darkest Africa." *Audubon,* July 1987.

Cooke, Deryck. *The Language of Music.* London: Oxford University Press, 1987.

Crosette, Barbara. "A Thai Monk Unlocks Song in the Earth." *The New York Times,* December 30, 1987.

Grant, Brian. *The Silent Ear: Deafness in Literature.* New York: Faber and Faber, 1988.

Mach, Elyse, ed. *Great Pianists Speak for Themselves.* 2 vols. New York: Dodd, Mead & Co., 1988.

Rothman, Tony, and Amy Mereson. "Fiddling with the Future." *Discover,* September 1987.

Schaeffer, R. Murray. *The Composer in the Classroom.* Toronto: Clark and Cruickshank, 1965.

Schonberg, Harold. *Facing the Music.* New York: Summit Books, 1985.

"School in the Exploratorium Idea Sheets." San Francisco: The Exploratorium Bookstore, n.d.

VISION

Bataille, Georges, trans. J. Neugroschal. *Story of the Eye.* San Francisco: City Lights Books, 1987.

————, trans. Allen Stockl. *Visions of Excess: Selected Writings 1927–1939.* Minneapolis: University of Minneapolis Press, 1985.

Berger, John. *About Looking.* New York: Pantheon Books, 1980.

————. *The Sense of Sight.* New York: Pantheon Books, 1980.

Bova, Ben. *The Beauty of Light.* New York: John Wiley & Sons, Inc., 1988.

Koretz, Jane F., and George H. Handelman. "How the Human Eye Focuses." *Scientific American,* July 1988.

Merleau-Ponty, Maurice, trans. H. L. and T. A. Dreyfus. *Sense and Non-Sense.* Evanston, Illinois: Northwestern University Press, 1964.

Rossotti, Hazel. *Colour: Why the World Isn't Grey.* Princeton, New Jersey: Princeton University Press, 1983.

Shearer, Lloyd. "A Doctor Who Advertises." *Parade,* July 24, 1988.

Taylor, Joshua C. *Learning to Look: A Handbook for the Visual Arts.* Chicago, Illinois: University of Chicago Press, 1957.

Trevor-Roper, Patrick. *The World Through Blunted Sight.* London: Penguin Books, 1988.

Vaughan, Christopher. "A New View of Vision." *Science News,* July 23, 1988.

INDEX

ABOUT THE AUTHOR

DIANE ACKERMAN was born in Waukegan, Illinois. She received her B.A. in English from Pennsylvania State University, and an M.F.A. and Ph.D. from Cornell University. She is the author of five collections of poems, *The Planets: A Cosmic Pastoral* (1976), *Wife of Light* (1978), *Lady Faustus* (1983), *Reverse Thunder: A Dramatic Poem* (1988), and *Jaguar of Sweet Laughter: New and Selected Poems* (1991). Her books of nonfiction include *Twilight of the Tenderfoot* (1980), *On Extended Wings* (1985), and *The Moon by Whale Light* (1991). She has received the Academy of American Poets' Peter I. B. Lavan Award, and grants from the National Endowment for the Arts and the Rockefeller Foundation, among other prizes and awards. Ms. Ackerman has taught at a variety of universities, including Washington University, New York University, Columbia, and Cornell. She is currently a staff writer at *The New Yorker*, and lives in upstate New York.